Writers on W

To Jean, with love

Writers on Writing

An Anthology
Chosen by Robert Neale

Auckland
OXFORD UNIVERSITY PRESS
Oxford   New York   Toronto   Melbourne

Oxford University Press

Oxford New York
Athens Auckland Bangkok Bombay
Cacutta Cape Town Dar es Salaam Delhi
Florence Hong Kong Istanbul Karachi
Kuala Lumpur Madras Madrid Melbourne
Mexico City Nairobi Paris Singapore
Taipei Toyoko Toronto
and associated companies in
Berlin Ibadan

Oxford is a trade mark of Oxford University Press

First published 1992
Introduction, notes and selection © Robert Neale 1992
Reprinted 1995

National Library of New Zealand
Cataloguing-in-Publication data

Writers on writing: an anthology/chosen by Robert Neale.
  Auckland, N.Z.: Oxford University Press, 1992.
  1v.
  Includes bibliographical references and index.
  1. English language—Writing. 2. English language—Rhetoric. 3. Authorship.
I. Neale, Robert, 1993-
808.042 (808.8)

ISBN 0 19 558256 X

Cover designed by Nik Andrew
Typeset in Adobe Galliard by Egan-Reid Ltd
Printed in New Zealand by GP Print Ltd
Published by Oxford University Press
540 Great South Road, Greenlane, Auckland 5, New Zealand

CONTENTS

INTRODUCTION

So here I am, in the middle way, having had twenty years —
Twenty years largely wasted, the years of *l'entre deux guerres* —
Trying to learn to use words, and every attempt
Is a wholly new start, and a different kind of failure
Because one has only learnt to get the better of words
For the thing one no longer has to say, or the way in which
One is no longer disposed to say it. And so each venture
Is a new beginning, a raid on the inarticulate
With shabby equipment always deteriorating
In the general mess of imprecision of feeling,
Undisciplined squads of emotion. And what there is to conquer
By strength and submission, has already been discovered
Once or twice, or several times, by men whom one cannot hope
To emulate — but there is no competition —
There is only the fight to recover what has been lost
And found and lost again and again: and now, under conditions
That seem unpropitious. But perhaps neither gain nor loss.
For us, there is only the trying. The rest is not our business.

T. S. Eliot, *East Coker* V

When you want to learn the tricks — and the trials — of a trade or craft you do well to listen to the advice of an expert; and before you have a go yourself you are generally wise to watch one at work. While the products of painters, musicians, carpenters, goldsmiths and so on certainly comment, however obliquely, on existing artefacts of the same sort, such craftspeople normally discuss their work and analyse their skills in a totally different medium — language. Writers alone produce their artefacts in the very medium in which they discuss them; which means that anyone interested in learning the writer's craft can very conveniently listen to what the experts have to say while simultaneously watching them at work. Writers, while their pens remain in their hands, are never, as it were, off duty. Even people with the temerity to write introductions to collections of others' writing run more than the normal risk of finding their own efforts under scrutiny. Whatever deficiencies such scrutiny is bound to reveal, this book seeks to help aspiring writers of all kinds by collecting together some of the things writers of acknowledged achievement have written on the subject of their own craft.

Few would dispute the value of doing so. The world today acknowledges — more widely perhaps than ever before — the political, economic, and cultural importance of effective writing, increasingly recognizing it as a craft that people can and should teach and learn. In response to this recognition, scholars, teachers, and researchers continue to look for better ways of managing and evaluating such teaching; and, inevitably, handbooks on how to write proliferate.

A surprising number of these, by a perverse irony, are themselves shockingly badly written, dreary in detail, and forbiddingly unreadable as a whole. Some, perhaps half-conscious of their inadequacy, claim to be not so much for consecutive study as for reference: identify your problem, and the book, properly searched, will provide a solution — a claim somewhat in conflict with the usual experience that, in writing, identifying a problem goes more than half way towards solving it, and that in any case

> every attempt
> Is a wholly new start, and a different kind of failure.

But no matter how assiduously its readers search it, no writing handbook can ever teach them to write; each venture founders, in practice, on the same rocky fact — that the only way of learning to write is to *write*: you can no more learn to write by merely reading about it than you can learn to swim, or weave, by reading about that. At best, such books tabulate some of the handier conventions of the writer's craft — techniques for 'pre-writing' and proofreading, the uses of the semi-colon, the dangers of the dangling participle, the tricks of handling parallel construction, and so on. That they continue to pour forth merely accumulates evidence of their inadequacy as a species, while reminding us that good teachers can make a silk purse out of almost anything while bad ones will grasp at any straw and somehow convert it into a crutch.

The teaching of writing must always depend on the skill of the teacher. Education, as P. G. Wodehouse somewhere reminds us, is a drawing out, not a putting in. This, though no doubt grossly oversimplifying a complex process, applies perhaps more to the business of writing than to any other part of learning. Writing, to be any good at all, must call up something already in the writer, taking outward shape in response to some inward pressure. Students may, for various reasons, acquire a certain superficial fluency, but will commit themselves wholeheartedly to their writing only if it matters to them. The grade at the foot of the exercise seems to matter only fleetingly to most students, while conscientious pedagogic comments and annotations usually matter hardly at all. Humane and capable teachers can, on the other hand, identify and set in motion certain other considerations which students will take very seriously indeed.

We all, in the first place, matter enormously to ourselves, and will bother about our writing if we feel ourselves to be somehow at stake or on show as we write. We can never, of course, display at any one time the entire complex of characteristics that makes us what we are. Writing of different sorts brings different parts of our personalities into play: a job-application will present a different *persona* from that revealed in a declaration of love, while any subsequent letters of resignation (or to 'Dear John') will reflect yet further facets of our elusive essence. But there must be something of me in everything I write, otherwise my writing degenerates either into plagiarism or into callous and deliberate insincerity. We have traditionally trained some of our students to be very good at both of these.

Equally, writers must presume — or at least pretend to presume — that someone wants to read them. Northrop Frye defines the academic research thesis as 'a document which is, practically by definition, something that nobody particularly wants either to write or to read' — a statement which, if true, goes a long way towards damning the whole system. It may be more true of the humanities than of the sciences, and reflect the way in which twentieth-century economic and cultural pressures have forced the former to model themselves upon the latter, to eschew their real function in favour of often factitious 'research,' and to prefer quantity to quality. Professor Frye has something to say about that, too — and it may well help to account for the proliferation of writers' handbooks. Academic essay-writing, at any level and whatever its merits, essentially amounts to telling your teacher or tutor what that person must be presumed already to know. Teaching your grandparent thus to suck eggs will hardly inspire you to go to very much trouble. A good teacher will understand that if I am to take my writing seriously I need to feel that it will matter to someone else.

Then, for writing to communicate, it must follow, or deliberately flout, the conventions of some genre or combination of genres. Herein lies a tension: that which lurks within the writer — personal, unique, and probably with little real shape, 'the general mess of imprecision of feeling' — emerges into the world to find a variety of prefabricated structures, of varying degrees of suitability, waiting to receive and mould it. These structures range from individual words to great traditional genres like epic and tragedy. We cannot avoid or ignore them: they are all we have, and in their various ways play an essential part in conveying meaning. But no word or genre will probably fit, precisely and absolutely, whatever new motion of the mind slouches towards Bethlehem to be born: modifications become necessary on both sides, with new and unexpected meanings often emerging in the process. Thus the genre of science fiction liberates and reinvigorates Doris Lessing's creative imagination. Writers revise and rework in order to reach some kind of compromise between the demands of what they think they want to say and the pressures of the available genres — although often all that most of us can claim is that one way of putting it 'feels' better than another. We have to trust in instinct based on experience.

Experience, of course, helps: a writer familiar with various genres and their conventions will stand a better chance of guessing right. But such familiarity alone will not generate good writing, which binds matter and manner, 'inspiration' (for want of a better word) and convention, into a complete, indivisible, and powerful whole. Poetry inevitably provides the readiest examples: Sir Philip Sidney finds a traditional sonnet (slightly modified) the best form for expressing his passionate feelings about *Loving in Truth*; Wordsworth, setting out to praise the sonnet form, has no choice but to use it; e. e. cummings, suspicious of 'the syntax of things,' rejects (in part) the syntax of language; and other poets represented in this book embody their ideas in whatever verse form — couplet, stanzaic, free, or blank — dovetails best with them. John Betjeman both utters and illustrates the point:

> Verse seems to me the shortest way
> Of saying what one has to say.

Prose too has its genres; to this book Samuel Johnson contributes one kind of biography and James Boswell another, while George Borrow, Rudyard Kipling and Arthur Koestler bring their various versions of autobiography. Edward Gibbon and Mary Augusta Ward produce sub-species of this latter in the form of the memoir (a form picked up for similar purposes in the twentieth century by George Orwell, Janet Frame, and Witi Ihimaera), while from Katherine Mansfield and Virginia Woolf come versions (edited by their respective husbands) of that most intimate autobiographical genre, the journal. Writers as diverse in every way as William Caxton, Samuel Johnson and Doris Lessing provide examples of the preface, and Francis Bacon, John Locke, Jonathan Swift, William Hazlitt and George Orwell explore the flexible and accommodating genre of the essay. Robert Frost, Northrop Frye and Lauris Edmond all give spoken addresses, a genre which, when printed, necessarily poses special problems: several things in Lauris Edmond's *Imagining Ourselves* needed modification as the demands of print replaced those of the microphone. Milton's *Areopagitica*, though designated 'a Speech' in its title, was intended not for reading aloud but for distribution as a polemical pamphlet. In *The Decay of Lying* Oscar Wilde exploits the dialogue, a form that dates back at least to Plato and to the Book of Job. Each piece of writing sits firmly in a generic tradition, but each piece in its turn modifies and redirects that tradition: Boswell leaves as deep a mark on biography as does Shakespeare on the sonnet.

Each genre thus not only dictates its own requirements, but also characteristically reinforces and reinterprets the subject-matter that suits it. Subject-matter inevitably preoccupies most writers in the first place: unless writing entirely at someone else's behest, we write because we have something to write about, something inside ourselves, *something to say*. This too no handbook can impart, and this too a good teacher can discover and encourage. Who can tell how we acquire the enthusiasms and interests that form part of the infinite complexity and mystery of our lives? Certainly talking helps: we hear other people talk about their experiences, and explore our own by articulating them into speech. In practice, talking builds a halfway house between having the experience and writing about it; by talking, we can test and, if necessary, discard whole versions of our experience — attitudes to it, points of view upon it, valuations of it. We make ourselves laugh or cry, hope or despair, reject or accept, deny or affirm, all by virtue of the words we find. As the Red Queen very reasonably points out to Alice during her adventures through the looking-glass: 'Once you've said a thing, that fixes it, and you must take the consequences.' Or, in W. H. Auden's words, 'The Ogre cannot master speech.' Language *makes* life, in the end; and 'putting a thing into words' (inadequate phrase!) can make it matter to us enough to make us want to write about it.

The foregoing four categories — the writer's *persona*, readership, genre, and subject-matter — have enjoyed increasing recognition in recent years as the four essential variables in any act of writing. Collected under the acronym RAFT (for **R**ole, **A**udience, **F**ormat, and **T**opic) they provide a buoyant platform on which to float our ideas out into the luxury and peril of public scrutiny. Handbooks on writing, if they recognize them at all, can have very little to say in detail about any of them, because each arises very specifically from the circumstances of the moment. Of the four, genre lends itself most readily to systematic analysis, and some very good studies exist. These offer more value to readers than to writers, however, since a proper understanding of what we read demands a proper grasp of its genre: most detective novels, for example, become unreadable if we ignore the basic assumptions and limitations of the genre and demand too much psychological consistency and depth in their characters. As writers we can acquire a more useful understanding of genre and its types by reading and studying examples rather than by merely reading *about* them. Any writer, that is to say, lives and works amid a whole lot of other writers rather than among a set of printed rules and systematized conventions. When we write we have at our back everything we have ever read. We are wise, therefore, to read the best.

Which, to recapitulate, justifies this book. It focuses directly upon the business of writing, collecting together discussions thereon, in prose and verse, by writers of acknowledged and lasting repute. These are the experts. What they say, therefore, will well repay close attention. So will the way they say it. Their opinions, preoccupations and priorities vary widely, of course;

so, indeed, does their practice, as even a quick survey of the book will show. A selection ranging over two-and-a-half millennia renders this variety inevitable, but there seems little consistency even among those writers — the great majority in the book — alive in the last two or three hundred years. Some, like Ezra Pound, emphasize the discipline of the writing process itself; others concentrate on one or more of a variety of theoretical or background issues.

All, however, share one overriding concern — to work out how best to tell the truth in writing. This, the concern of all writers except pulp novelists, advertising copy-writers, a few depraved journalists, and (George Orwell would claim) politicians, further justifies this book. A preoccupation with the truth, perhaps with the Truth, denotes the human race and distinguishes it from all other species on this planet; and this preoccupation emerges in its writing: all the great religions of the world focus upon the Scriptures wherein their followers seek to find the Truth expressed in one way or another. This Truth may or may not be simple; the way in which words conceal or announce it is certainly very complex indeed. All the extracts in this book reveal their authors' fascination with the resources available to them for expressing the truths they have discovered, or with the process of writing itself as a means of discovering the truth.

Most of us begin life with a naïvely literal view of how language reflects reality. The figurative properties of words regularly baffle children; my younger daughter to this day recalls her perplexity when I once suggested to her that she should 'put herself in X's shoes' (I disown the cliché, but she insists). John Locke, wrestling with questions of mimesis in the extract from his *Essay Concerning Human Understanding*, argues with rigorous and unassailable logic that words express the ideas in our heads rather than the things around us — that, indeed, no hard or fast connection can exist between words and objects, otherwise there could be only one language. To supplement Locke's argument we may note, with Sir Francis Bacon, the various ways in which various forms of writing handle language's essential separateness. The Chinese may do it differently, but in the western tradition pictogrammatic forms have largely given way to alphabetic: our writing, which began by drawing *pictures of things*, has increasingly come to concentrate on reproducing the *sounds of words*.

Language, written or spoken, thus works at some distance from any 'objective reality' that we may deem to exist, and if it is to 'tell the truth' it must acknowledge this distance, even to the extent of accepting the paradox that the only way to tell the truth is to tell lies. Sir Philip Sidney suggests this, with the proviso that a writer whose audience expects a fiction can never be accused of lying anyway. Oscar Wilde characteristically goes a step further, maintaining the moral admirability of such lying and lamenting its current decay. What is truth? Can language, he implies, ever convey anything other than some kind of fiction, needing all our skill to control it if it is to encompass any truth at all?

Such questions about the mimetic function of language have exercised writers from Aristotle onwards. His laconic insights, preserved in the document known as *Poetics* or *The Art of Poetry*, prompt more detailed discussions like those of Sidney and Wilde and that by Sir Francis Bacon in his *Advancement of Learning*. And notions of *mimesis* animate practically every word written by the writers represented in this book, from Milton's defence of authorial freedom to Jane Austen's championship of the novel's reputation, from Dr Johnson's lexicographical sonorities to the raw and merciless self-criticism of Katherine Mansfield's diary. Writing, they all assume, struggles to convey truth — whether factual, social or political truths of the world around, or emotional or psychological truths of the personality within.

And it very often does so through metaphor. Aristotle again first perceives the centrality of metaphor in our attempts to make sense of the world, and later writers inevitably rehearse the idea. Dr Johnson, in his *Life of Cowley*, defines for all time that special and self-conscious species of metaphor, the metaphysical conceit, as

> . . . a kind of *discordia concors*, a combination of dissimilar images, or discovery of occult resemblances in things apparently unlike. . . . The most heterogeneous ideas . . . yoked by violence together. . . .

Yet the metaphysical conceit surely possesses no monopoly of such heterogeneity; Johnson himself, clarifying his point by means of the word 'yoked', paints a complex little picture of two animals of different species harnessed uncomfortably together to the plough, 'things apparently most unlike' the figures of speech he is talking about.

Metaphor lives at the very heart of language, and we have to live with its complexities and ambiguities. Its fullest and most profound treatment in this book comes from American poet Robert Frost, whose conclusions about the educational importance of metaphor (and hence of poetry, its most potent manifestation) may prompt us to think further about how it shapes our view of the world we live in and of the humanity we belong to. Pre-literate cultures enshrined their values and their very identity in their orally transmitted mythology. As, over the last three thousand years, writing has spread worldwide, so books and documents have become the depositories of our mythologies (using the word in its broadest sense). The world's great religions, as we have seen, centre upon their Scriptures; and documents like Marx's *Das Kapital*, the American Declaration of Independence, the

Thoughts of Chairman Mao and the Treaty of Waitangi (among innumerable others) have played their part in shaping humanity's view of itself. All, of course, are made of words, not things: all, therefore, function in some way metaphorically and so stand in need of interpretation.

Any metaphor compares two apparently unlike things; indeed, metaphor's strength lies in its paradoxical and disturbing claim that for all their dissimilarity the two are basically identical. Scholars call one of these things — the one under discussion, that we are already aware of — the *tenor*, and the new thing (introduced by way of comparison) the *vehicle*. Thus T. S. Eliot uses military terminology as the vehicle for conveying his struggles with words, while in Dr Johnson's metaphor just quoted the 'heterogeneous ideas' form the tenor, while the plough-animals constitute the vehicle. We can, if we wish, stick to our childhood assumptions about language and try to read such a statement literally; we may well ignore the animals altogether — many readers would indeed simply not notice the metaphorical force of the word 'yoked.' Some such process works constantly to drain the colour from vigorous metaphors and bleach them into cliché. The extremely literal-minded, on the other hand, might consider bringing a large wooden yoke on which to inscribe (violently) their 'heterogeneous ideas.' Totalitarians and fundamentalists of all kinds, it seems to me, share this naïve approach to language, denying the ambiguity that inheres in anything made of words and insisting on the necessity both of a literal reading and of only one 'correct' meaning. Totalitarianism thus produces 'the party line,' secret police, and inevitable dictatorship, while fundamentalism — since no two people interpret any bit of language in quite the same way — generates schism; both rely heavily upon slogans and formulae, all of which, while current, are deemed to possess some absolute and unambiguous truthfulness. My childhood memories include church services where the preacher urged us not to sing the announced hymn unless we 'really meant' its words. I, puzzled hypocrite, sang anyway.

To counter such misconceptions we can, as Frost suggests, educate ourselves in an awareness of metaphor. Though unlikely to solve all our problems, this will at least remind us that any set of words may have more than one viable interpretation — and even endow us with the humility to accept the limitations of our own views. The greatest metaphors of all, enshrining the deepest truths, of course remain inscrutable; no interpretation (i.e. no alternative set of words) can express more than a tiny fraction of their full meaning. The Bible has generated millions of sermons and treatises, but continues to transcend them all: no amount of exegesis will ever fully account for the Parable of the Good Samaritan. In the beginning was indeed the Word, and humanity can never finally get around or behind it. Our grasp of truth begins and ends with metaphor. And in the things of deepest import, at most we apprehend the *vehicle*: the *tenor* remains for ever beyond our knowing. In the words of Robert Frost's own finely balanced and metaphorical epigram:

> We dance round in a ring and suppose,
> But the Secret sits in the middle and knows.

In the light of all this it is hardly surprising that, in their struggle to express the truth, few of the writers represented in this book limit themselves merely to questions of formal correctness — the kind of concern that first presses on most of us as would-be writers and keeps 'the dictionary' constantly at our elbow today. No book of rules can contain the answers to the problems of language, although anything to do with communication must rely heavily on some agreed system of conventions. Thus Chaucer, naturally enough, expresses irritation with the copying errors of his scribe Adam. William Caxton, a century later, faces the problem on a much larger scale; he has watched the English language pass through some of the fastest and most thoroughgoing changes in its history, and has lived through a time when grandparents were finding their grandchildren's utterances even harder to understand than usual. And having your breakfast order misunderstood — whether for 'eggys' or 'eyren' — matters at any time. So to Caxton the choice of the right word presents a real dilemma. His instincts, however, lead him already to see the problem as one of appropriateness rather than of correctness — and this emphasis guides most of the other writers in the book who think about the problem. Jonathan Swift, who should perhaps have known better, voices the most authoritarian view in his *Tatler* article — though we may sympathize with his disapproval of the pretentiousness and obscurity of some of the slang he is attacking. We may even conclude that framing an anonymous letter in a pseudonymous article effectively exonerates Swift from the very opinions he appears to be endorsing.

Dr Johnson has most to say about correct English; his views are worth extremely careful perusal, and should not be distorted by the reputation for authority that has grown up around him and around 'the dictionary' whose prestige he did so much to establish. To his own lexicographical *Preface* he brings a balanced judgement and unrivalled experience of the language. As a result he, like Caxton before him, sees it as a working tool with a communicative job to do. The excerpt from James Boswell's *Life of Johnson* shows the great man taking much the same liberal position on matters of etymology and pronunciation. Otherwise only William Hazlitt, advocating 'the familiar style', explores the topic of correctness, and though ostensibly disapproving of everything Johnson stands for, inevitably also finds himself in agreement with the principle that we judge a writer's 'correctness' not by any absolute canons but by reference to such issues as *persona*, audience, genre, and subject-matter — to the whole RAFT, in fact.

The writing process itself provides perhaps the single most arresting topic of this book. Insights into its problems abound — from Geoffrey Chaucer's frustrations over his copyist's incompetence, through John Milton's blindness, to Lauris Edmond's maternal and domestic preoccupations. Such problems, of course, defy generalization; they vary from age to age and from individual to individual; we may simply find reassurance that they have been so resoundingly surmounted, usually by means of such rigorous self-disciplines as those suggested by Ezra Pound in his *ABC of Reading* or those implied in many of the poems printed in this book (see, for example, those by Sidney, Pope, Cowper, Blake, Wordsworth, Elizabeth Barrett Browning, Hopkins, Edward Thomas, Stephen Spender, and Ted Hughes). The evidence seems overwhelmingly to suggest that the writers represented here have generally achieved their eminence in spite of rather than because of their environments. Life itself, of course, throws up an interminable series of distractions from any kind of creativity or consistency.

Nevertheless while some writers, particularly those who focus directly upon the topics we have already discussed, merely imply the problems they have faced (though each supplies plenty of evidence for the alert reader), the experience so lovingly and painfully recounted by others like George Borrow, Mary Augusta Ward, Rudyard Kipling, Virginia Woolf, Katherine Mansfield, George Orwell, Janet Frame, and Witi Ihimaera, suggests an inextricable interrelationship between a writer's difficulties and the nature and quality of the writing that emerges: Robert Herrick finds 'this dull Devonshire' surprisingly conducive to poetry. The struggle from the chrysalis forms the butterfly; or, as Lauris Edmond puts it, 'Once I say to myself that life is one thing, art another, I believe I do irreparable damage to both. I have learnt not to try.'

May all aspiring writers so learn. To underestimate the difficulty and complexity of language, to deny the essential ambiguity of its metaphoric nature, to subject it to a set of arbitrary rules, is to try to separate it from the life that it reflects, embodies, and indeed largely shapes. This holds true whether or not we set out to write 'creatively' — indeed, once we begin seriously to investigate how language works, we will find any fundamental distinction between 'creative' and 'non-creative' writing increasingly difficult to sustain. The writer, whether of epics or essays, odes or office-memoranda, 'trying to learn to use words,' struggling to pin down some of this protean element for the information, enjoyment, or edification of others, may well share the pessimism voiced by the greatest English-language poet of the twentieth century in the passage quoted at the head of this Introduction. But for us all, as for Eliot, 'there is only the trying. The rest is not our business.'

According to George Orwell, 'writing a book is a horrible, exhausting struggle, like a long bout of some painful illness.' This holds true, in some measure, even of an anthology such as this, and I feel much gratitude to the many people who have helped ease its passage: particularly to Anne French and Oxford University Press for encouragement, fertile suggestions, and financial support; to Marcela Sierra Contreras, who put the whole thing on disk and provided tactful but invaluable criticism; to Dr Simon Cauchi, without whose meticulous scholarship the book would lack even more than it does; to Massey University, for providing facilities and equipment and for underwriting some of the costs; and above all to my wife, for faith, hope, and charity beyond measure. To her the book is dedicated, with all love.

Robert Neale
Massey University, New Zealand

ACKNOWLEDGEMENT

The editor thanks the Publication Committee of Massey University, Palmerston North, for generously supporting this work with a publication grant.

A Note on the Texts

In any book on writing we do well to pay some attention to the way in which a piece of writing takes visual shape as well as to that shape as it finally meets our eyes. This too concerns writers: Rudyard Kipling demands black ink, Virginia Woolf laments 'I must correct the re-re-typing,' and so on. Reprinting each piece in this book has imposed the usual conflicting demands: a typeface, page-layout and context that the piece has never previously enjoyed (or perhaps suffered) necessarily give it a new shape and presumably take it another stage further from what its author originally envisaged and wrote (not to speak of any proofreading errors that may have escaped my vigilance). But in spite of that we would all like what is printed here to remain as faithful as possible to its original.

That, however, is a tall order: the majority of the authors printed here wrote their work in manuscript form, with a pen, using letter-shapes and orthographic conventions impossible or inconvenient to reproduce in print — which can never hope to replicate, for example, the curving dashes with which Emily Dickinson punctuated her poems. Chaucer and his copyists of course accommodated their script and punctuation to the inconveniences and limitations of the parchment on which they wrote. Print, furthermore, developed its own conventions, which were sometimes very far removed from the conventions of manuscript. Many of these manuscripts, moreover, no longer exist; we have to rely upon a series of printed versions, each one edited and unavoidably modified just as I have had to edit and modify. Of these, the earliest is not necessarily the most reliable. Modern scholars constantly apply their increasingly sophisticated research-tools to producing 'definitive editions' of the works of writers of the past. But the conventions of printing, like those of handwriting, change with time, demanding from editors a perpetual compromise between what the original author and reader would have expected and what suits a modern audience and press. (Aristotle and Chaucer, of course, lived before the printing-press was invented — though the twentieth-century translation of the former that we print here, while short-circuiting some problems, increases others enormously if we really want to find out what the author actually wrote!)

My choice of authors and excerpts (from the vast resources available) of course reflects my own tastes, preferences, and limitations. Similarly in choosing source-texts I have deliberately taken pot luck. These are listed and acknowledged in another place; most are reliable modern scholarly editions of the 'definitive' type mentioned above. In a few cases, however, I have turned to older versions: the extract from Caxton's *Prologue to Eneydos,* for example, follows the Early English Text Society's edition which tries to convey some of the qualities of the text Caxton printed and issued himself.

And Sir Philip Sidney's *Defence of Poesie* reproduces some of the printing conventions of Sidney's own time, themselves derived from the habits of the old manuscript scribes. These things are dealt with in the Notes. Different editors, moreover, may work according to very different principles: some will regularize, or modernize, their material far more than others. A glance at the two extracts from the writings of Francis Bacon will suggest some of the decisions editors have to take, and the variety of solutions possible. As usual, no 'correct' way of handling all these problems presents itself; like all other aspects of writing, they demand a balanced compromise which takes into account the RAFT of variables discussed in the Introduction above.

I and Pangur Bán my cat,
'Tis a like task we are at:
Hunting mice is his delight,
Hunting words I sit all night.

Better far than praise of men
'Tis to sit with book and pen;
Pangur bears me no ill will,
He too plies his simple skill.

'Tis a merry thing to see
At our tasks how glad are we,
When at home we sit and find
Entertainment to our mind.

Oftentimes a mouse will stray
In the hero Pangur's way;
Oftentimes my keen thought set
Takes a meaning in its net.

'Gainst the wall he sets his eye
Full and fierce and sharp and sly;
'Gainst the wall of knowledge I
All my little wisdom try.

When a mouse darts from his den
O how glad is Pangur then!
O what gladness do I prove
When I solve the doubts I love!

So in peace our tasks we ply,
Pangur Bán, my cat, and I;
In our arts we find our bliss,
I have mine and he has his.

Practice every day has made
Pangur perfect in his trade;
I get wisdom day and night
Turning darkness into light.

Anon.[1] (9c. Gaelic, transl. Robin Flower)

ARISTOTLE

The fragment known as the *Poetics* by the Greek philosopher Aristotle (384–322 B.C.) represents a very small section of his extant writings, themselves only a modest fraction of the huge corpus of philosophical and scientific work he is believed to have produced. It has nevertheless exercised an enormous influence on subsequent writers, and therefore on the cultures they have helped shape. The extracts reproduced here, in Ingram Bywater's translation, embody Aristotle's central doctrine of Imitation or Mimesis and his famous statement on Metaphor.

from *Poetics*

I t is clear that the general origin of poetry was due to two causes, each of them part of human nature. Imitation is natural to man from childhood, one of his advantages over the lower animals being this, that he is the most imitative creature in the world, and learns at first by imitation. And it is also natural for all to delight in works of imitation. The truth of this second point is shown by experience: though the objects themselves may be painful to see, we delight to view the most realistic representations of them in art, the forms for example of the lowest animals and of dead bodies. The explanation is to be found in a further fact: to be learning something is the greatest of pleasures not only to the philosopher but also to the rest of mankind, however small their capacity for it; the reason of the delight in seeing the picture is that one is at the same time learning — gathering the meaning of things, e.g. that the man there is so-and-so; for if one has not seen the thing before, one's pleasure will not be in the picture as an imitation of it, but will be due to the execution or colouring or some similar cause. Imitation, then, being natural to us — as also the sense of harmony and rhythm, the metres being obviously species of rhythms — it was through their original aptitude, and by a series of improvements for the most part gradual on their first efforts, that they created poetry out of their improvisations. . . .

The perfection of Diction is for it to be at once clear and not mean. The clearest indeed is that made up of ordinary words for things, but it is mean, as is shown by the poetry of Cleophon and Sthenelus.[1] On the other hand the Diction becomes distinguished and non-prosaic by the use of unfamiliar terms, i.e. strange words, metaphors, lengthened forms, and everything that deviates from the ordinary modes of speech. — But a whole statement in such terms will either be a riddle or a barbarism, a riddle, if made up of metaphors, a barbarism, if made up of strange words. The very nature indeed of a riddle is this, to describe a fact in an impossible combination of words (which cannot be done with the real names for things, but can be with their metaphorical substitutes); e.g. 'I saw a man glue brass on another with fire', and the like. The corresponding use of strange words results in a barbarism.— A certain admixture, accordingly, of unfamiliar terms is necessary. These, the strange word, the metaphor, the ornamental equivalent, etc., will save the language from seeming mean and prosaic, while the ordinary words in it will secure the requisite clearness. What helps most, however, to render the Diction at once clear and non-prosaic is the use of the lengthened, curtailed, and altered forms of words. Their deviation from the ordinary words will, by making the language unlike that in general use, give it a non-prosaic appearance; and their having much in common with the words in general use will give it the quality of clearness. . . .

It is a great thing, indeed, to make a proper use of these poetical forms, as also of compounds and strange words. But the greatest thing by far is to be a master of metaphor. It is the one thing that cannot be learnt from others; and it is also a sign of genius, since a good metaphor implies an intuitive perception of the similarity in dissimilars.

Chaucers Wordes unto Adam, His Owne Scriveyn [1]

Adam scriveyn, if ever it thee bifalle
Boece or *Troylus* for to wryten newe,[2]
Under thy long lokkes thou most have the scalle,
But after my makyng thow wryte more trewe;[3]
So ofte a-daye I mot thy werk renewe
It to correcte and eek to rubbe and scrape;[4]
And al is thorugh thy negligence and rape.

Geoffrey Chaucer (*ca.* 1340–1400)

WILLIAM CAXTON

William Caxton (*ca.* 1421–91) introduced the printing press into England and exercised a powerful influence on national life through his choice of works for publication. His life spanned a period of extremely rapid linguistic change in English, generating for the writer the kinds of problem Caxton discusses in this prefatory statement to his translation, published in 1490, of a French version of the Roman poet Vergil's epic poem *The Aeneid*. The difficulties of fifteenth-century spelling are compounded by the interchangeability of the letters v and u. Some of the harder words are explained in the endnotes. Caxton makes frequent use of the slash (/) as a punctuation device.

from *Eneydos: The Prologue*

After dyuerse werkes made / translated and achieued / hauyng noo werke in hande, I, sittyng in my studye where as laye many dyuerse paunflettis and bookys, happened that to my hande came a lytyl book in frenshe, whiche late was translated oute of latyn by some noble clerke of fraunce, which booke is named Eneydos / made in latyn by that noble poete & grete clerke vyrgyle / whiche booke I sawe ouer and redde therin, How, after the generall destruccyon of the grete Troye, Eneas departed, berynge his olde fader anchises vpon his sholdres / his lityl son yolus on his honde, his wyfe wyth moche other people folowynge / and how he shypped and departed, wyth alle thystorye of his aduentures that he had er he cam to the achieuement of his conquest of ytalye, as all a longe shall be shewed in this present boke. In whiche booke I had grete playsyr, by cause of the fayr and honest termes & wordes in frenshe / whyche I neuer sawe to-fore lyke, ne none so playsaunt ne so wel ordred; whiche booke, as me semed, sholde be moche requysyte to noble men to see, as wel for the eloquence as the historyes / How wel that many honderd yerys passed was the sayd booke of eneydos, wyth other werkes, made and lerned dayly in scolis, specyally in ytalye & other places / whiche historye the sayd vyrgyle made in metre / And whan I had aduysed me in this sayd boke, I delybered and concluded to translate it in-to englysshe, And forthwyth toke a penne & ynke, and wrote a leef or tweyne / whyche I ouersawe agayn to corecte it / And whan I sawe the fayr & straunge termes therin / I doubted that it sholde not please some gentylmen whiche late blamed me, sayeng *that* in my translacyons I had ouer curyous termes whiche coude not be vnderstande of comyn peple / and desired me to vse olde and homely termes in my translacyons. and / fayn wolde I satysfye euery man / and so to doo, toke an olde boke and redde therin / and certaynly the englysshe was so rude and brood that I coude not wele vnderstande it. And also my lorde abbot of westmynster ded do shewe to me late, certayn euydences[1] wryton in olde englysshe, for to reduce it in-to our englysshe now vsid / And certaynly it was wreton in suche wyse that it was more lyke to dutche[2] than englysshe; I coude not reduce ne brynge it to be vnderstonden / And certaynly our langage now vsed varyeth ferre from that whiche was vsed and spoken whan I was borne / For we englysshe men / ben borne vnder the domynacyon of the mone, whiche is neuer stedfaste / but euer wauerynge / wexynge one season / and waneth & dyscreaseth another season / And that comyn englysshe that is spoken in one shyre varyeth from a nother. In so moche that in my dayes happened that certayn merchauntes were in a shippe in tamyse,[3] for to haue sayled ouer the see into zelande / and for lacke of wynde, thei taryed atte forlond, and wente to lande for to refreshe them; And one of theym named sheffelde, a mercer, cam in-to an hows[4] and axed for mete; and specyally he axyd after eggys; And the goode wyf anwerde, that she coude speke no frenshe. And the marchaunt was angry, for he also coude speke no frenshe, but wolde haue hadde egges / and she vnderstode hym not / And thenne at laste a nother sayd that he wolde haue eyren / then the good wyf sayd that she vnderstod hym wel / Loo, what sholde a man in thyse dayes now wryte, egges or eyren / certaynly it is harde to playse euery man / by cause of dyuersite & chaunge of langage. For in these dayes euery man that is in ony reputacyon in his countre, wyll vtter his commynycacyon and maters in suche maners & termes / that fewe men shall vnderstonde theym / And som honest and grete clerkes haue ben wyth me, and desired me to wryte the moste curyous termes that I coude fynde / And thus betwene playn rude / & curyous, I stande abasshed. but in my Iudgemente / the comyn termes that be dayli vsed, ben lyghter to be vnderstonde than the olde and auncyent englysshe / And for as moche as this present booke is not for a rude vplondyssh[5] man to laboure therin / ne rede it / but onely for a clerke & a noble gentylman that feleth and vnderstondeth in faytes of armes, in loue, & in noble chyualrye / Therfor in a meane bytwene bothe, I haue reduced & translated this sayd booke in to our englysshe, not ouer rude ne curyous, but in suche termes as shall be vnderstanden, by goddys grace, accordynge to my copye. And yf ony man wyll enter-mete[6] in redyng of hit, and fyndeth suche termes that he can not vnderstande, late hym goo rede and lerne vyrgyll / or the pystles of ouyde[7] / and ther he shall see and vnderstonde lyghtly all / Yf he haue a good redar & enformer / For this booke is not for euery rude and / vnconnynge[8] man to see / but to clerkys and very gentylmen that vnderstande gentylnes and scyence.[9] Thenne I praye alle theym that shall rede in this lytyl treatys, to holde me for excused for the translatynge of hit. For I knowleche my selfe ignorant of connynge to enpryse on me so hie and noble a werke / But I praye mayster Iohn Skelton,[10] late created poet laureate in the vnyuersite of oxenforde, to ouersee and correcte this sayd booke, And taddresse and expowne where as shalle be founde faulte to theym that shall requyre it. For hym, I knowe for suffycyent to expowne and englysshe euery dyffyculte that is therin / For he hath late translated the epystlys of Tulle / and the boke of dyodorus syculus, and diuerse other werkes oute of latyn in-to englysshe, not in rude and olde langage, but in polysshed and ornate termes craftely,[11] as he that hath redde vyrgyle / ouyde, tullye, and all the other noble poetes and oratours / to me vnknowen: And also he hath redde the ix. muses, and vnderstande theyr musicalle scyences, and to whom of theym eche scyence is appropred. I suppose he hath dronken of Elycons well. Then I praye hym, & suche other, to correcte, adde or mynysshe where as he or they shall fynde faulte / For I haue but folowed my copye in frenshe as nygh as me is possyble / And yf ony worde be sayd therin well / I am glad; and yf otherwyse, I submytte my sayd boke to theyr correctyon / Whiche boke I presente vnto the hye born my tocomynge[12] naturell and souerayn lord, Arthur, by the grace of god, Prynce of Walys, Duc of Cornewayll, & Erle of Chester, fyrst begoten sone and heyer vnto our most dradde naturall & souerayn lorde, & most crysten kynge / Henry the vij. by the grace of god, kynge of Englonde and of Fraunce, & lord of Irelonde / byseching his noble grace to receyue it in thanke of me, his most humble subget & seruaunt / And I shall praye vnto almyghty god for his prosperous encreasyng in vertue / wysedom / and humanyte, that he may be egal wyth the nost renommed of alle his noble progenytours. And so to lyue in this present lyf / that after this transitorye lyfe he and we alle may come to euerlastynge lyf in heuen / Amen:

from *Astrophel and Stella*

Loving in truth, and faine in verse my love to show,
That she (deare she) might take some pleasure of my paine:
Pleasure might cause her reade, reading might make her know,
Knowledge might pitie winne, and pitie grace obtaine,
   I sought fit words to paint the blackest face of woe,
Studying inventions fine, her wits to entertaine:
Oft turning others' leaves, to see if thence would flow
Some fresh and fruitfull showers upon my sunne-burn'd braine.
   But words came halting forth, wanting Invention's stay,
Invention, Nature's child, fled step-dame Studie's blowes,
And others' feete still seem'd but strangers in my way.
Thus great with child to speake, and helplesse in my throwes,
   Biting my trewand pen, beating my selfe for spite,
   'Foole,' said my Muse to me, 'looke in thy heart and write.'

Sir Philip Sidney (1554–86)

SIR PHILIP SIDNEY

Sir Philip Sidney (1554–86) by the age of twenty was on intimate terms with most of the leading intellectual and political figures of Europe. His prose and poetry were widely read, and he died memorably in a skirmish in Flanders. *The Defence of Poesie* (otherwise known as *An Apologie for Poetry*) articulates his reaction to a contemporary Puritan attack on imaginative literature of all kinds, and incorporates some development of Aristotle's ideas about the mimetic function of writing. The version printed here follows that published by William Ponsonby in 1594.

from *The Defence of Poesie*

But now, let us see how the Greekes have named it, and how they deemed of it. The Greekes named him ποιητήν,[1] which name, hath as the most excellent, gone through other languages, it commeth of this word ποιεῖν[2] which is to make: wherin I know not whether by luck or wisedom, we Englishmen have met with the Greekes in calling him a Maker. Which name, how high and incomparable a title it is, I had rather were knowne by marking the scope of other sciences, thē[3] by any partial allegatiō. There is no Art delivered unto mankind that hath not the workes of nature for his principall object, without which they could not consist, and on which they so depend, as they become Actors and Plaiers, as it were of what nature will have set forth. So doth the *Astronomer* looke upon the starres, and by that he seeth set downe what order nature hath taken therein. So doth the *Geometritian* & *Arithmititian*, in their divers sorts of quantities. So doth the *Musitians* in times tel you, which by nature agree, which not. The natural *Philosopher* thereon hath his name, and the morall *Philosopher* standeth upon the naturall vertues, vices, or passions of man: and follow nature saith he therein, and thou shalt not erre. The *Lawier* saith, what men have determined. The *Historian*, what men have done. The *Gramarian*, speaketh onely of the rules of speech, and the *Rhetoritian* and *Logitian*, considering what in nature wil soonest proove, and perswade thereon, give artificiall rules, which still are compassed within the circle of a question, according to the proposed matter. The *Phisitian* wayeth the nature of mans bodie, & the nature of things helpfull, or hurtfull unto it. And the *Metaphisicke* though it be in the second & abstract Notions, and therefore be counted supernaturall, yet doth hee indeed build upon the depth of nature. Onely the Poet disdeining to be tied to any such subjectiō, lifted up with the vigor of his own invention, doth grow in effect into an other nature: in making things either better then nature bringeth foorth, or quite a new, formes such as never were in nature: as the *Heroes, Demigods, Cyclops, Chymeras, Furies,* and such like; so as he goeth hand in hand with nature, not enclosed within the narrow warrant of her gifts, but freely raunging within the Zodiack of his owne wit. Nature never set foorth the earth in so rich Tapistry as diverse Poets have done, neither with so pleasaunt rivers, fruitfull trees, sweete smelling flowers, nor whatsoever els may make the too much loved earth more lovely: her world is brasen, the Poets only deliver a golden. But let those things alone and goe to man, for whom as the other things are, so it seemeth in him her uttermost comming[4] is imploied: & know whether she have brought foorth so true a lover as *Theagenes,* so constant a friend as *Pylades,* so valiant a man as *Orlando,* so right a Prince as *Xenophons Cyrus,* so excellent a man every way as *Virgils Aeneas.*[5] Neither let this be jestingly cōceived, bicause the works of the one be essenciall, the other in imitation or fiction: for everie understanding, knoweth the skill of ech Artificer standeth in that *Idea,* or fore conceit of the worke, and not in the worke it selfe. And that the Poet hath that *Idea,* is manifest, by delivering them foorth in such excellencie as he had imagined them: which delivering foorth, also is not wholly imaginative, as we are wont to say by thē that build Castles in the aire: but so farre substancially it worketh, not onely to make a *Cyrus,* which had bene but a particular excellency as nature might have done, but to bestow a *Cyrus* upon the world to make many *Cyrusses,* if they will learne aright, why and how that maker made him. Neither let it be deemed too sawcy a comparison, to ballance the highest point of mans wit, with the efficacie of nature: but rather give right honour to the heavenly maker of that maker, who having made man to his owne likenes, set him beyond and over all the workes of that second nature, which in nothing he sheweth so much as in Poetry; when with the force of a divine breath, he bringeth things foorth surpassing her doings: with no small arguments to the incredulous of that first accursed fall of *Adam,* since our erected wit maketh us know what perfectiō is, and yet our infected wil keepeth us frō reaching unto it. But these argumēts will by few be understood, and by fewer graunted: thus much I hope wil be given me, that the Greeks with some probability of reason, gave him the name above all names of learning. Now let us goe to a more ordinarie opening of him, that the truth may be the more palpable: and so I hope though we get not so unmatched a praise as the *Etimologie* of his names will graunt, yet his verie description which no man will denie, shall not justly be barred from a principall commendation. *Poesie* therefore, is an Art of *Imitation:* for so *Aristotle* termeth it in the word μιμησις,[6] that is to say, a representing, counterfeiting, or figuring forth to speake Metaphorically. A speaking *Picture,* with this end to teach and delight. . . .

Now therein of all Sciences I speake still of humane (and according to the humane conceit) is our *Poet* the *Monarch.* For hee doth not onely shew the way, but giveth so sweete a prospect into the way, as will entice anie man to enter into it: Nay he doth as if your journey should lye through a faire vineyard, at the verie first, give you a cluster of grapes, that full of that taste, you may long to passe further. Hee beginneth not with obscure definitions, which must blurre the margent with interpretations, and loade the memorie with doubtfulnesse: but hee commeth to you with words set in delightfull proportion, either accompanied with, or prepared for the well enchanting skill of *Musicke*, and with a tale forsooth he commeth unto you, with a tale, which holdeth children from play, and olde men from the Chimney corner; and pretending no more, doth intend the winning of the minde from wickednes to vertue; even as the child is often brought to take most wholesome things by hiding them in such other as have a pleasaunt taste: which if one should begin to tell them the nature of the *Aloes* or *Rhabarbarum* they should receive, wold sooner take their phisick at their eares then at their mouth, so is it in men (most of which, are childish in the best things, til they be cradled in their graves) glad they will be to heare the tales of *Hercules, Achilles, Cyrus, Aeneas,* and hearing them, must needes heare the right description of wisdom, value and justice; which if they had bene barely (that is to say Philosophically) set out, they would sweare they be brought to schoole againe; that imitation whereof *Poetrie* is, hath the most conveniencie to nature of al other: insomuch that as *Aristotle* saith, those things which in themselves are horrible, as cruel battailes, unnatural monsters, are made in poeticall imitation, delightfull. Truly I have knowne men, that even with reading *Amadis de gaule,*[7] which God knoweth, wanteth much of a perfect *Poesie,* have found their hearts moved to the exercise of courtesie, liberalitie, and especially courage. Who readeth *Aeneas* carrying old *Anchises* on his backe, that wisheth not it were his fortune to performe so excellent an Act? Whom doth not those words of *Turnus* moove, (the Tale of *Turnus* having planted his image in the imagination) *fugientem hæc terra videbit? Usqueadeone mori miserum est?*[8] Wher the *Philosophers* as they think scorne to delight, so must they be content little to moove: saving wrangling whether *Virtus* be the chiefe or the onely good; whether the contemplative or the active life do excell; which *Plato & Boetius* well knew: and therefore made mistresse *Philosophie* verie often borrow the masking raiment of *Poesie.*[9] For even those hard hearted evill men who thinke vertue a schoole name, and know no other good but *indulgere genio,*[10] and therefore despise the austere admonitions of the *Philosopher,* and feele not the inward reason they stand upon, yet will be content to be delighted, which is al the good, fellow *Poet* seemes to promise; and so steale to see the form of goodnes, (which seene, they cannot but love) ere themselves be aware, as if they tooke a medicine of Cheries. . . .

Since then *Poetrie* is of al humane learnings the most ancient, and of most fatherly antiquitie, as from whence other learnings have taken their beginnings; Since it is so universall, that no learned nation doth despise it, nor barbarous nation is without it; Since both *Romane & Greeke* gave such divine names unto it, the one of prophesying, the other of making; and that indeed that name of making is fit for him, considering, that where all other Arts retain themselves within their subject, and receive as it were their being from it. The *Poet* onely, onely bringeth his own stuffe, and doth not learn a Conceit out of a matter, but maketh matter for a Conceit. Since neither his description, nor end, containing any evill, the thing described cannot be evil; since his effects be so good as to teache goodnes, and delight the learners of it; since therein (namely in morall doctrine the chiefe of all knowledges) hee doth not onely farre passe the *Historian*, but for instructing is well nigh comparable to the *Philosopher*, for moving, leaveth him behind him. Since the holy scripture (wherein there is no uncleannesse) hath whole parts in it Poeticall, and that even our Savior Christ vouchsafed to use the flowers of it: since all his kindes are not onely in their united formes, but in their severed dissections fully commendable, I thinke, (and thinke I thinke rightly) the Lawrell Crowne appointed for tryumphant Captaines, doth worthily of all other learnings, honour the *Poets* triumph. But bicause we have eares aswell as toongs, and that the lightest reasons that may be, will seeme to waigh greatly, if nothing be put in the counterballance, let us heare, and as well as we can, ponder what objections be made against this Art, which may be woorthie either of yeelding, or answering. . . .

Now then goe we to the most important imputations laid to the poore *Poets,* for ought I can yet learne, they are these. First, that there beeing manie other more frutefull knowledges, a man might better spend his time in them, then in this. Secondly, that it is the mother of lyes. Thirdly, that it is the nurse of abuse, infecting us with many pestilent desires, with a *Sirens* sweetnesse, drawing the minde to the Serpents taile of sinfull fansies; and herein especially *Comedies* give the largest field to eare, as *Chawcer* saith,[11] how both in other nations and in ours, before *Poets* did soften us, we were full of courage givē to martial exercises, the pillers of manlike libertie, and not lulled a sleepe in shadie idlenes, with *Poets* pastimes. And lastly and chiefly, they cry out with open mouth as if they had overshot *Robinhood*, that *Plato* banished them out of his Commonwealth. Truly this is much, if there be much truth in it. First to the first. That a man might better spend his time, is a reason indeed: but it doth as they say, but *petere principium*.[12] For if it be, as I affirme, that no learning is so good, as that which teacheth and moveth to vertue, and that none can both teach and move thereto so much as *Poesie*, then is the conclusion manifest; that incke and paper cannot be to a more profitable purpose imployed. And certainly though a man should graunt their first assumption, it should follow (mee thinks) very unwillingly, that good is not good, because better is better. But I still and utterly deny, that there is sprung out of earth a more fruitfull knowledge. To the second therefore, that they should be the principall lyers, I answere *Paradoxically*, but truly, I think truly: that of all writers under the Sunne, the *Poet* is the least lyer: and though he wold, as a *Poet* can scarcely be a lyer. The *Astronomer* with his cousin the *Geometrician*, can hardly escape, when they take upon them to measure the height of the starres. How often thinke you do the *Phisitians* lie, when they averre things good for sicknesses, which afterwards send *Charon*[13] a great number of soules drownd in a potion, before they come to his Ferrie? And no lesse of the rest, which take upon them to affirme. Now for the *Poet*, he nothing affirmeth, and therefore never lieth: for as I take it, to lie, is to affirme that to bee true, which is false. So as the other *Artistes,* and especially the *Historian,* affirming manie things, can in the clowdie knowledge of mankinde, hardly escape from manie lies. But the *Poet* as I said before, never affirmeth, the *Poet* never maketh any Circles about your imaginatiō, to conjure you to beleeve for true, what he writeth: he citeth not authorities of other histories, but evē for his entrie, calleth the sweete *Muses* to inspire unto him a good invention. In troth, not laboring to tel you what is, or is not, but what should, or should not be. And therefore though he recount things not true, yet because he telleth them not for true, he lieth not: without we will say, that *Nathan,* lied in his speech before alleaged to *David,*[14] which as a wicked man durst scarce say, so think I none so simple wold say, that *Esope* lied in the tales of his beasts: for who thinketh that *Esope* wrote it for actually true, were wel worthie to have his name Cronicled among the beasts he writeth of. What childe is there, that coming to a play, and seeing *Thebes* written in great letters upon an old doore, doth beleeve that it is *Thebes?* If then a man can arrive to the childes age, to know that the *Poets* persons and dooings, are but pictures, what should be, and not stories what have bin, they will never give the lie to things not Affirmatively, but Allegorically and figuratively written; and therefore as in historie looking for truth, they may go away full fraught with falshood: So in *Poesie,* looking but for fiction, they shal use the narration but as an imaginative groundplat of a profitable invention. But hereto is replied, that the *Poets* give names to men they write of, which argueth a conceit of an actuall truth, and so not being true, prooveth a falshood. And dooth the *Lawier* lye, then when under the names of *John* of the *Stile* and *John* of the *Nokes,* hee putteth his Case? But that is easily answered, their naming of men, is but to make their picture the more lively, and not to build anie Historie. Painting men, they cannot leave men namelesse: wee see, wee cannot plaie at Chestes, but that wee must give names to our Chessemen; and yet mee thinkes he were a verie partiall Champion of truth, that would say wee lyed, for giving a peece of wood the reverende title of a Bishop. The *Poet* nameth *Cyrus* and *Aeneas,* no other way, then to shewe what men of their fames, fortunes, and estates, should doo. Their third is, how much it abuseth mens wit, training it to wanton sinfulnesse, and lustfull love. For indeed that is the principall if not onely abuse, I can heare alleadged. They say the *Comedies* rather teach then reprehend amorous cōceits. They say the *Lirick* is larded with passionat

Sonets, the *Elegiack* weeps the want of his mistresse, and that even to the *Heroical, Cupid* hath ambitiously climed. Alas Love, I would thou couldest as wel defende thy selfe, as thou canst offend others: I would those on whom thou doest attend, could either put thee away, or yeeld good reason why they keepe thee. But grant love of bewtie to be a beastly fault, although it be verie hard, since onely man and no beast hath that gift to discerne bewty, graunt that lovely name of love, to deserve all hatefull reproches, although even some of my maisters the *Philosophers* spent a good deale of their Lampoyle in setting foorth the excellencie of it, graunt I say, what they will have graunted, that not onelie love, but lust, but vanitie, but if they list scurrilitie, possesse manie leaves of the *Poets* bookes, yet thinke I, when this is graunted, they will finde their sentence may with good manners put the last words foremost; and not say, that *Poetrie* abuseth mans wit, but that mans wit abuseth *Poetrie.* For I will not denie, but that mans wit may make *Poesie,* which should be εικαστικη,[15] which some learned have defined figuring foorth good things to be φαυταστικη:[16] which doth contrariwise infect the fancie with unwoorthy objects, as the Painter should give to the eye eyther some excellent perspective, or some fine Picture fit for building or fortification, or containing in it some notable example, as *Abraham* sacrificing his son *Isaack, Judith* killing *Holofernes, David* fighting with *Golias,* may leave those, and please an ill pleased eye with wanton shewes of better hiddẽ matters. But what, shal the abuse of a thing, make the right use odious? Nay truly, though I yeeld, that *Poesie* may not onely be abused, but that being abused by the reason of his sweete charming force, it can do more hurt then anie other armie of words: yet shall it be so farre from concluding, that the abuse should give reproach to the abused, that cõtrariwise, it is a good reason, that whatsoever being abused, doth most harme, being rightly used (and upon the right use, ech thing receives his title) doth most good. Do we not see skill of Phisicke the best ramper[17] to our often assaulted bodies, being abused, teach poyson the most violent destroyer? Doth not knowledge of Law, whose end is, to even & right all things, being abused, grow the crooked fosterer of horrible injuries? Doth not (to go to the highest) Gods word abused, breede heresie, and his name abused, become blasphemie? Truly a Needle cannot do much hurt, and as truly (with leave of Ladies be it spoken) it cannot do much good. With a swoord thou maist kill thy Father, and with a swoord thou maist defend thy Prince and Countrey: so that, as in their calling *Poets,* fathers of lies, they said nothing, so in this their argument of abuse, they proove the commendation. They alledge herewith, that before *Poets* began to be in price, our Nation had set their hearts delight uppon action, and not imagination, rather doing things worthie to be written, thẽ writing things fit to be done. What that before time was, I think scarcely *Spinx* can tell: since no memerie is so ancient, that hath not the precedens of *Poetrie.* And certain it is, that in our plainest homelines yet never was the *Albion* Nation without *Poetrie*. Marry this Argument, though it be leviled against *Poetrie*, yet is it indeed, a chain-shot against all learning or bookishnes, as they commonly terme it. Of such mind were certaine *Gothes*, of whom it is written, that having in the spoile of a famous Cittie, taken a faire Librarie, one hangman belike fit to execute the frutes of their wits, who had murthered a great number of bodies, woulde have set fire in it. No said another verie gravely, take heed what you do, for while they are busie about those toyes, wee shall with more leisure conquere their Countries. This indeed is the ordinarie doctrine of ignorance, and many words sometimes I have heard spent in it: but bicause this reason is generally against all learning, as wel as *Poetrie*, or rather all learning but *Poetrie*, because it were too large a digression to handle it, or at least too superfluous, since it is manifest that all government of action is to be gotten by knowledge, and knowledge best, by gathering manie knowledges, which is reading; I onely with *Horace*, to him that is of that opinion, *Jubio stultum esse libenter:* [18] for as for *Poetrie* it selfe, it is the freest from this objection, for *Poetrie* is the Companion of Camps. I dare undertake, *Orlando Furioso*, or honest king *Arthure*, will never displease a souldier: but the quidditie of *Ens & Prima materia*, [19] will hardly agree with a Corcelet. And therefore as I said in the beginning, even *Turkes* and *Tartars*, are delighted with *Poets*. *Homer* a *Greeke*, flourished, before *Greece* flourished: and if to a slight conjecture, a conjecture may bee apposed, truly it may seem, that as by him their learned mẽ tooke almost their first light of knowledge, so their active men, received their first motions of courage. Onely *Alexanders* example may serve, who by *Plutarche* is accounted of such vertue, that fortune was not his guide, but his footstoole, whose Acts speake for him, though *Plutarche* did not: indeede the *Phœnix* of warlike Princes. This *Alexander*, left his Schoolemaister living *Aristotle* behinde him, but tooke dead *Homer* with him. Hee put the Philosopher *Callisthenes* to death, for his seeming Philosophicall, indeed mutinous stubbornnesse, but the chiefe thing hee was ever heard to wish for, was, that *Homer* had bene alive. Hee well founde hee received more braverie of minde by the paterne of *Achilles*, then by hearing the definition of fortitude. And therefore if *Cato* misliked *Fulvius* for carrying *Ennius* with him to the field, It may be answered, that if *Cato* misliked it, the Noble *Fulvius* liked it, or else he had not done it; for it was not the excellent *Cato Uticencis*, whose authoritie I would much more have reverenced: But it was the former, in truth a bitter punisher of faultes, but else a man that had never sacrificed to the *Graces*. Hee misliked and cried out against all Greeke learning, and yet being foure score yeares olde beganne to learne it, belike fearing that *Pluto* understoode not Latine. Indeed the *Romane* lawes allowed no person to bee carried to the warres, but hee that was in the souldiers Role. And therefore though *Cato* misliked his unmustred person, he misliked not his worke. And if hee had, *Scipio Nasica* (judged by common consent the best *Romane*) loved him: both the other *Scipio* brothers, who had by their vertues no lesse surnames then of *Asia* and *Affricke*, so loved him, that they caused his bodie to be buried in their Sepulture. So as *Catoes* authoritie beeing but against his person, and that answered with so farre greater then himselfe, is herein of no validitie. But now indeede my burthen is great, that *Plato* his name is laide uppon mee, whom I must confesse of all *Philosophers*, I have ever esteemed most worthie of reverence; and with good reason, since of all *Philosophers* hee is the most *Poeticall.* . . .

Plato therefore, whose authoritie, I had much rather justly cōsture,[20] then unjustly resist: ment not in generall of *Poets*, in those words of which *Julius Scaliger* saith: *Qua authoritate barbari quidam, atque hispidi abuti velint ad poetas è rep. exigendos.*[21] But only ment to drive out those wrong opinions of the Deitie: wherof now without further law, *Christianitie* hath taken away all the hurtful beliefe, perchance as he thought nourished by then esteemed *Poets*. And a man need go no further then to *Plato* himselfe to knowe his meaning: who in his Dialogue called *Ion*, giveth high, and rightly divine commendation unto *Poetrie*. So as *Plato* banishing the abuse, not the thing, not banishing it, but giving due honour to it, shall be our Patron, and not our adversarie. For indeed, I had much rather, since truly I may do it, shew their mistaking of *Plato*, under whose Lyons skinne, they would make an Aslike braying against *Poesie*, then go about to overthrow his authoritie; whome the wiser a man is, the more just cause he shall finde to have in admiration: especially since he attributeth unto *Poesie*, more then my selfe do; namely, to be a verie inspiring of a divine force, farre above mans wit, as in the forenamed Dialogue is apparant. Of the other side, who would shew the honours have bene by the best sort of judgements graunted them, a whole sea of examples would present themselves; *Alexanders*, *Cæsars*, *Scipioes*, all favourers of *Poets*. *Lælius*, called the Romane *Socrates* himselfe a *Poet*; so as part of *Heautontimorumenon* in *Terence*, was supposed to bee made by him. And even the Greeke *Socrates*, whome *Appollo* confirmed to bee the onely wise man, is said to have spent part of his olde time in putting *Esopes* Fables into verses. And therefore full evill should it become his scholler *Plato*, to put such words in his maisters mouth against *Poets*. But what needs more? *Aristotle* writes the Arte of *Poesie*, and why, if it should not bee written? *Plutarche* teacheth the use to bee gathered of them, and how, if they should not bee reade? And who reades *Plutarches* either Historie or *Philosophie*, shall finde hee trimmeth both their garments with gardes of *Poesie*. But I list not to defend *Poesie* with the helpe of his underling *Historiographie*. Let it suffice to have shewed, it is a fit soyle for praise to dwell uppon: and what dispraise may set uppon it, is either easily overcome, or transformed into just commendation. So that since the excellencies of it, may bee so easily and so justly confirmed, and the lowe creeping objections so soone trodden downe, it not beeing an Art of lyes but of true doctrine;

not of effœminatenesse, but of notable stirring of courage; not of abusing mans wit, but of strengthening mans wit; not banished, but honored by *Plato*: Let us rather plant more Lawrels for to ingarland the *Poets* heads (which honor of being Lawreate, as besides them onely triumphant Captaines were, is a sufficient authoritie to shewe the price they ought to bee held in) then suffer the ill savoured breath of such wrong speakers once to blow uppon the cleare springs of *Poesie*. . . .

But what? methinks, I deserve to be poūded for straying from *Poetrie*, to *Oratory*: but both have such an affinitie in the wordish consideratiõ, that I think this digression will make my meaning receive the fuller understanding: which is not to take upon me to teach *Poets* how they should do, but only finding my selfe sicke among the rest, to shew some one or two spots of the common infection growne among the most part of writers; that acknowledging our selves somewhat awry, wee may bende to the right use both of matter and manner. Whereto our language giveth us great occasion, being indeed capable of any excellent exercising of it. I knowe some will say it is a mingled language: And why not, so much the better, taking the best of both the other? Another will say, it wanteth Grammer. Nay truly it hath that praise that it wants not Grammer; for Grammer it might have, but it needs it not, being so easie in it selfe, and so voyd of those combersome differences of *Cases, Genders, Moods*, & *Tenses*, which I thinke was a peece of the Tower of *Babilons* curse, that a man should be put to schoole to learn his mother tongue. But for the uttering sweetly and properly the conceit of the minde, which is the end of speech, that hath it equally with any other tongue in the world. And is perticularly happy in compositions of two or three wordes togither, neare the Greeke, farre beyond the Latine, which is one of the greatest bewties can be in a language.

from *Amoretti*

One day I wrote her name upon the strand;
But came the waves, and washéd it away:
Agayne, I wrote it with a second hand;
But came the tyde, and made my paynes his pray.
Vayne man, sayd she, that doest in vaine assay
A mortall thing so to immortalize;
For I my selve shall lyke to this decay,
And eek my name bee wyped out lykewize.
Not so, quod I; let baser things devize
To dy in dust, but you shall live by fame:
My verse your vertues rare shall éternize,
And in the hevens wryte your glorious name.
   Where, whenas death shall all the world subdew,
   Our love shall live, and later life renew.

Edmund Spenser (*ca.* 1552–99)

FRANCIS BACON

Francis Bacon, Lord Verulam (1561–1626) became Lord Chancellor of England in 1618 until his disgrace in 1621. Besides his well-known *Essays*, he wrote on law, philosophy, and education, and contributed largely to the rise of 'the scientific method' in his *Novum Organum*. Some people think he wrote the plays of Shakespeare. In these extracts from *The Advancement of Learning* he analyses some of the problems faced by Elizabethan writers and discusses language as a means of imitating experience.

from *The Advancement of Learning*

There be therefore chiefly three vanities in studies, whereby learning hath been most traduced. For those things we do esteem vain, which are either false or frivolous, those which either have no truth or no use; and those persons we esteem vain, which are either credulous or curious; and curiosity is either in matter or words: so that in reason as well as in experience there fall out to be these three distempers (as I may term them) of learning; the first, fantastical learning; the second, contentious learning; and the last, delicate learning; vain imaginations, vain altercations, and vain affectations; and with the last I will begin. Martin Luther, conducted (no doubt) by an higher providence, but in discourse of reason, finding what a province he had undertaken against the bishop of Rome and the degenerate traditions of the church, and finding his own solitude, being no ways aided by the opinions of his own time, was enforced to awake all antiquity and to call former times to his succours to make a party against the present time: so that the ancient authors, both in divinity and in humanity, which had long time slept in libraries, began generally to be read and revolved. This by consequence did draw on a necessity of a more exquisite travail in the languages original, wherein those authors did write, for the better understanding of those authors, and the better advantage of pressing and applying their words. And thereof grew again a delight in their manner of style and phrase, and an admiration of that kind of writing; which was much furthered and precipitated by the enmity and opposition that the propounders of those primitive but seeming new opinions had against the schoolmen; who were generally of the contrary part, and whose writings were altogether in a different style and form; taking liberty to coin and frame new terms of art to express their own sense, and to avoid circuit of speech, without regard to the pureness, pleasantness, and (as I may call it) lawfulness of the phrase or word. And again, because the great labour then was with the people (of whom the Pharisees were wont to say, *Execrabilis ista turba, quae non novit legem*),[1] for the winning and persuading of them there grew of necessity in chief price and request eloquence and variety of discourse, as the fittest and forciblest access into the capacity of the vulgar sort: so that these four causes concurring, the admiration of ancient authors, the hate of the schoolmen, the exact study of languages, and the efficacy of preaching, did bring in an affectionate study of eloquence and copie of speech, which then began to flourish. This grew speedily to an excess; for men began to hunt more after words than matter; more after the choiceness of the phrase, and the round and clean composition of the sentence, and the sweet falling of the clauses, and the varying and illustration of their works with tropes and figures, than after the weight of matter, worth of subject, soundness of argument, life of invention, or depth of judgement. Then grew the flowing and watery vein of Osorius, the Portugal bishop, to be in price. Then did Sturmius spend such infinite and curious pains upon Cicero the Orator, and Hermogenes the Rhetorician, besides his own books of Periods and Imitation, and the like. Then did Car of Cambridge and Ascham with their lectures and writings almost deify Cicero and Demosthenes, and allure all young men that were studious unto that delicate and polished kind of learning. Then did Erasmus take occasion to make the scoffing echo, *Decem annos consumpsi in legendo Cicerone*; and the echo answered in Greek *One, Asine*.[2] Then grew the learning of the schoolmen to be utterly despised as barbarous. In sum, the whole inclination and bent of those times was rather towards copie than weight.

Here therefore is the first distemper of learning, when men study words and not matter; whereof, though I have represented an example of late times, yet it hath been and will be *secundum majus et minus*[3] in all time. And how is it possible but this should have an operation to discredit learning, even with vulgar capacities, when they see learned men's works like the first letter of a patent, or limned book,[4] which though it hath large flourishes, yet it is but a letter? It seems to me that Pygmalion's frenzy is a good emblem or portraiture of this vanity: for words are but the images of matter; and except they have life of reason and invention, to fall in love with them is all one as to fall in love with a picture. . . .

There remaineth the fourth kind of rational knowledge, which is transitive,[5] concerning the expressing or transferring our knowledge to others; which I will term by the general name of tradition or delivery. Tradition hath three parts; the first concerning the organ of tradition; the second concerning the method of tradition; and the third concerning the illustration of tradition.

For the organ of tradition, it is either speech or writing: for Aristotle saith well, 'Words are the images of cogitations, and letters are the images of words.' But yet it is not of necessity that cogitations be expressed by the medium of words. For whatsoever is capable of sufficient differences, and those perceptible by the sense, is in nature competent to express cogitations. And therefore we see in the commerce of barbarous people, that understand not one another's language, and in the practice of divers that are dumb and deaf, that men's minds are expressed in gestures, though not exactly, yet to serve the turn. And we understand further, that it is the use of China, and the kingdoms of the High Levant, to write in characters real, which express neither letters nor words in gross, but things or notions; insomuch as countries and provinces, which understand not one another's language, can nevertheless read one another's writings, because the characters are accepted more generally than the languages do extend; and therefore they have a vast multitude of characters, as many (I suppose) as radical words.

De notis rerum.[6] These notes of cogitations are of two sorts; the one when the note hath some similitude or congruity with the notion: the other *ad placitum,*[7] having force only by contract or acceptation. Of the former sort are hieroglyphics and gestures. For as to hieroglyphics (things of ancient use, and embraced chiefly by the Egyptians, one of the most ancient nations), they are but as continued impreses and emblems.[8] And as for gestures, they are as transitory hieroglyphics, and are to hieroglyphics as words spoken are to words written, in that they abide not; but they have evermore, as well as the other, an affinity with the things signified. As Periander, being consulted with how to preserve a tyranny newly usurped, bid the messenger attend and report what he saw him do; and went into his garden and topped all the highest flowers: signifying, that it consisted in the cutting off and keeping low of the nobility and grandees. *Ad placitum*, are the characters real before mentioned, and words: although some have been willing by curious inquiry, or rather by apt feigning, to have derived imposition of names from reason and intendment; a speculation elegant, and, by reason it searcheth into antiquity, reverent; but sparingly mixed with truth, and of small fruit. This portion of knowledge, touching the notes of things, and cogitations in general, I find not inquired, but deficient. And although it may seem of no great use, considering that words and writings by letters do far excel all the other ways; yet because this part concerneth as it were the mint of knowledge (for words are the tokens current and accepted for conceits, as moneys are for values, and that it is fit men be not ignorant that moneys may be of another kind than gold and silver), I thought good to propound it to better inquiry.

Concerning speech and words, the consideration of them hath produced the science of grammar. For man still striveth to reintegrate himself in those benedictions, from which by his fault he hath been deprived; and as he hath striven against the first general curse by the invention of all other arts, so hath he sought to come forth of the second general curse (which was the confusion of tongues[9] by the art of grammar; wherof the use in a mother tongue is small, in a foreign tongue more; but most in such foreign tongues as have ceased to be vulgar tongues, and are turned only to learned tongues. The duty of it is of two natures: the one popular, which is for the speedy and perfect attaining of languages, as well for intercourse of speech as for understanding of authors; the other philosophical, examining the power and nature of words, as they are the footsteps and prints of reason: which kind of analogy between words and reason is handled *sparsim*,[10] brokenly though not entirely; and therefore I cannot report it deficient, though I think it very worthy to be reduced into a science by itself.

Unto grammar also belongeth, as an appendix, the consideration of the accidents of words; which are measure, sound, and elevation or accent, and the sweetness and harshness of them; whence hath issued some curious observations in rhetoric, but chiefly poesy, as we consider it, in respect of the verse and not of the argument. Wherein though men in learned tongues do tie themselves to the ancient measures, yet in modern languages it seemeth to me as free to make new measures of verses as of dances: for a dance is a measured pace, as a verse is a measured speech. In these things the sense is better judge than the art;

> Coenae fercula nostrae
> Mallem convivis quam placuisse cocis.[11]

And of the servile expressing antiquity in an unlike and an unfit subject, it is well said, *Quod tempore antiquum videtur, id incongruitate est maxime novum.*[12]

For ciphers,[13] they are commonly in letters, or alphabets, but may be in words. The kinds of ciphers (besides the simple ciphers, with changes, and intermixtures of nulls and non-significants) are many, according to the nature or rule of the infolding, wheel-ciphers, key-ciphers, doubles, &c. But the virtues of them, whereby they are to be preferred, are three; that they be not laborious to write and read; that they be impossible to decipher; and, in some cases, that they be without suspicion. The highest degree whereof is to write *omnia per omnia*,[14] which is undoubtedly possible, with a proportion quintuple at most of the writing infolding to the writing infolded, and no other restraint whatsoever. This art of ciphering hath for relative an art of deciphering, by supposition unprofitable, but, as things are, of great use. For suppose that ciphers were well managed, there be multitudes of them which exclude the decipherer. But in regard of the rawness and unskilfulness of the hands through which they pass, the greatest matters are many times carried in the weakest ciphers.

In the enumeration of these private and retired arts, it may be thought I seek to make a great muster-roll of sciences, naming them for show and ostentation, and to little other purpose. But let those which are skilful in them judge whether I bring them in only for appearance, or whether in that which I speak of them (though in few words) there be not some seed of proficience. And this must be remembered, that as there be many of great account in their countries and provinces, which, when they come up to the seat of the estate, are but of mean rank and scarcely regarded; so these arts, being here placed with the principal and supreme sciences, seem petty things; yet to such as have chosen them to spend their labours and studies in them, they seem great matters.

Sonnet 55

Not marble, nor the guilded monuments
Of Princes shall out-live this powrefull rime,
But you shall shine more bright in these contents
Then unswept stone, besmeer'd with sluttish time.
When wastefull warre shall *Statues* over-turne,
And broiles roote out the worke of masonry,
Nor *Mars* his sword, nor warres quick fire shall burne:
The living record of your memory.
Gainst death, and all oblivious enmity
Shall you pace forth, your praise shall stil finde roome,
Even in the eyes of all posterity
That weare this world out to the ending doome.
   So til the judgement that your selfe arise,
   You live in this, and dwell in lovers eies.

William Shakespeare (1564–1616)

FRANCIS BACON

One of Bacon's beautifully crafted little *Essays* discusses and illustrates what writing can achieve.

Of Studies

tudies serve for Delight, for Ornament, and for Ability. Their Chiefe Use for Delight, is in Privatenesse and Retiring; For Ornament, is in Discourse; And for Ability, is in the Judgement and Disposition of Businesse. For Expert Men can Execute, and perhaps Judge of particulars, one by one; But the generall Counsels, and the Plots, and Marshalling of Affaires, come best from those that are *Learned*. To spend too much Time in *Studies*, is Sloth; To use them too much for Ornament, is Affectation; To make Judgement wholly by their Rules is the Humour of a Scholler. They perfect Nature, and are perfected by Experience: For Naturall Abilities, are like Naturall Plants, that need Proyning by *Study*: And *Studies* themselves, doe give forth Directions too much at Large, except they be bounded in by experience. Crafty Men Contemne *Studies*; Simple Men Admire them; and Wise Men Use them: For they teach not their owne Use; But that is a Wisdome without them, and above them, won by Observation. Reade not to Contradict, and Confute; Nor to Beleeve and Take for granted; Nor to Finde Talke and Discourse; But to weigh and Consider. Some *Bookes* are to be Tasted, Others to be Swallowed, and Some Few to be Chewed and Digested: That is, some *Bookes* are to be read onely in Parts; Others to be read but not Curiously; And some Few to be read wholly, and with Diligence and Attention. Some *Bookes* also may be read by Deputy, and Extracts made of them by Others: But that would be, onely in the lesse important Arguments, and the Meaner Sort of *Bookes*: else distilled *Bookes*, are like Common distilled Waters, Flashy Things. Reading maketh a Full Man; Conference a Ready Man; And Writing an Exact Man. And therefore, If a Man Write little, he had need have a Great memory; If he Conferre little, he had need have a Present Wit; and if he Reade litle, he had need have much Cunning, to seeme to know that, he doth not. *Histories* make Men Wise; *Poets* Witty; The *Mathematicks* Subtill; *Natural Philosophy* deepe; *Morall* Grave; *Logick* and *Rhetorick* Able to Contend. *Abeunt studia in Mores.*[1] Nay there is no Stond or Impediment in the Wit, but may be wrought out by Fit *Studies*: Like as Diseases of the Body, may have Appropriate Exercises. Bowling is good for the Stone and Reines; Shooting for the Lungs and Breast; Gentle Walking for the Stomacke; Riding for the Head; And the like. So if a Mans Wit be

Wandring, let him *Study* the *Mathematicks*; For in Demonstrations, if his Wit be called away never so little, he must begin again: If his Wit be not Apt to distinguish or find differences, let him *Study* the *Schoole-men*; For they are *Cymini sectores*.[2] If he be not Apt to beat over Matters, and to call up one Thing, to Prove and Illustrate another, let him *Study* the *Lawyers Cases*: So every defect of the Minde, may have a Speciall Receit.

Discontents in Devon

More discontents I never had
    Since I was born, then here;
Where I have been, and still am sad,
    In this dull *Devon-shire:*
Yet justly too I must confesse;
    I ne'r invented such
Ennobled numbers for the Presse,
    Then where I loath'd so much.

Robert Herrick (1591–1674)

JOHN MILTON

John Milton (1608–74), poet, polemicist, and Latin (i.e. Foreign) Secretary to the
Council of State under Oliver Cromwell, wrote *Areopagitica: a Speech of Mr John
Milton for the Liberty of Unlicensed Printing to the Parliament of England* in protest
against a recent government order forbidding the publication of books and pamphlets without prior licensing. The title refers to the Areopagus or Hill of Ares, site
of democratic debate in ancient Athens. Milton's eventual blindness did not prevent
him from composing his great poems *Paradise Lost, Paradise Regained,* and *Samson
Agonistes.*

from *Areopagitica*

I deny not, but that it is of greatest concernment in the Church and
Commonwealth, to have a vigilant eye how Bookes demeane[1] themselves,
as well as men; and thereafter to confine, imprison, and do sharpest justice
on them as malefactors: For Books are not absolutely dead things, but doe
contain a potencie of life in them to be as active as that soule was whose
progeny they are; nay they do preserve as in a violl the purest efficacie and
extraction of that living intellect that bred them. I know they are as lively,
and as vigorously productive, as those fabulous Dragons teeth; and being
sown up and down, may chance to spring up armed men.[2] And yet on the
other hand unlesse warinesse be us'd, as good almost kill a Man as kill a good
Book; who kills a Man kills a reasonable creature, Gods Image; but hee who
destroyes a good Booke, kills reason it selfe, kills the Image of God, as it were
in the eye. Many a man lives a burden to the Earth; but a good Booke is the
pretious life-blood of a master spirit, imbalm'd and treasur'd up on purpose
to a life beyond life. 'Tis true, no age can restore a life, whereof perhaps there
is no great losse; and revolutions of ages doe not oft recover the losse of a
rejected truth, for the want of which whole Nations fare the worse. We
should be wary therefore what persecution we raise against the living labours
of publick men, how we spill that season'd life of man preserv'd and stor'd
up in Books; since we see a kinde of homicide may be thus committed,
sometimes a martyrdome, and if it extend to the whole impression, a kinde
of massacre, whereof the execution ends not in the slaying of an elementall
life, but strikes at that ethereall and fift essence,[3] the breath of reason it selfe,
slaies an immortality rather then a life. But lest I should be condemn'd of
introducing licence, while I oppose Licencing, I refuse not the paines to be
so much Historicall, as will serve to show what hath been done by ancient
and famous Commonwealths, against this disorder, till the very time that
this project of licencing crept out of the *Inquisition,*[4] was catcht up by our
Prelates, and hath caught some of our Presbyters. . . .

I conceive, therefore, that when God did enlarge the universall diet of mans body, saving ever the rules of temperance, he then also, as before, left arbitrary the dyeting and repasting of our minds; as wherein every mature man might have to exercise his owne leading capacity. How great a vertue is temperance, how much of moment through the whole life of man? yet God committs the managing so great a trust, without particular Law or prescription, wholly to the demeanour of every grown man. And therefore when he himself tabl'd the Jews from heaven, that Omer which was every mans daily portion of Manna,[5] is computed to have bin more then might have well suffic'd the heartiest feeder thrice as many meals. For those actions which enter into a man, rather then issue out of him, and therefore defile not,[6] God uses not to captivat under a perpetuall childhood of prescription, but trusts him with the gift of reason to be his own chooser; there were but little work left for preaching, if law and compulsion should grow so fast upon those things which hertofore were govern'd only by exhortation. *Salomon* informs us that much reading is a wearines to the flesh,[7] but neither he, nor other inspir'd author tells that such, or such reading is unlawfull: yet certainly had God thought good to limit us herein, it had bin much more expedient to have told us what was unlawfull, then what was wearisome. As for the burning of those Ephesian books by St *Pauls* converts, tis reply'd the books were magick, the Syriack so renders them.[8] It was a privat act, a voluntary act, and leaves us to a voluntary imitation: the men in remorse burnt those books which were their own; the Magistrat by this example is not appointed: these men practiz'd the books, another might perhaps have read them in some sort usefully. Good and evill we know in the field of this World grow up together almost inseparably; and the knowledge of good is so involv'd and interwoven with the knowledge of evill, and in so many cunning resemblances hardly to be discern'd, that those confused seeds which were impos'd on *Psyche* as an incessant labour to cull out, and sort asunder, were not more intermixt.[9] It was from out the rinde of one apple tasted, that the knowledge of good and evill as two twins cleaving together leapt forth into the World.[10] And perhaps this is that doom which *Adam* fell into of knowing good and evill, that is to say of knowing good by evill. As therefore the state of man now is; what wisdome can there be to choose, what continence to forbeare without the knowledge of evill? He that can apprehend and consider vice with all her baits and seeming pleasures, and yet abstain, and yet distinguish, and yet prefer that which is truly better, he is the true warfaring Christian. I cannot praise a fugitive and cloister'd vertue, unexercis'd & unbreath'd, that never sallies out and sees her adversary, but slinks out of the race, where that immortall garland is to be run for, not without dust and heat. Assuredly we bring not innocence into the world, we bring impurity much rather: that which purifies us is triall, and triall is by what is contrary. That vertue therefore which is but a youngling in the contemplation of evill, and knows not the utmost that vice promises to her followers, and rejects it, is but a blank vertue, not a pure; her whiteness is but an excrementall[11] whitenesse; Which was the reason why our sage and serious Poet *Spencer*, whom I dare be known to think a better teacher then *Scotus* or *Aquinas*,[12] describing true temperance under the person of *Guion*, brings him in with his palmer through the cave of Mammon,[13] and the bowr of earthly blisse that he might see and know, and yet abstain. Since therefore the knowledge and survay of vice is in this world so necessary to the constituting of human vertue, and the scanning of error to the confirmation of truth, how can we more safely, and with lesse danger scout into the regions of sin and falsity then by reading all manner of tractats, and hearing all manner of reason? And this is the benefit which may be had of books promiscuously read. . . .

Another reason, whereby to make it plain that this order will misse the end it seeks, consider by the quality which ought to be in every licencer. It cannot be deny'd but that he who is made judge to sit upon the birth, or death of books whether they may be wafted into this world, or not, had need to be a man above the common measure, both studious, learned and judicious; there may be else no mean mistakes in the censure of what is passable or not; which is also no mean injury. If he be of such worth as behoovs him, there cannot be a more tedious and unpleasing journey-work, a greater losse of time levied upon his head, then to be made the perpetuall reader of unchosen books and pamphlets, oftimes huge volumes. There is no book that is acceptable unlesse at certain seasons; but to be enjoyn'd the reading of that at all times, and in a hand scars legible, whereof three pages would not down at any time in the fairest Print, is an imposition which I cannot beleeve how he that values time, and his own studies, or is but of a sensible nostrill should be able to endure. In this one thing I crave leave of the present licencers to be pardon'd for so thinking: who doubtlesse took this office up, looking on it through their obedience to the Parlament, whose command perhaps made all things seem easie and unlaborious to them; but that this short triall hath wearied them out already, their own expressions and excuses to them who make so many journeys to sollicit their licence, are testimony anough. Seeing therefore those who now possesse the imployment, by all evident signs wish themselves well ridd of it, and that no man of worth, none that is not a plain unthrift of his own hours is ever likely to succeed them, except he mean to put himself to the salary of a Presse-corrector, we may easily foresee what kind of licencers we are to expect hereafter, either ignorant, imperious, and remisse, or basely pecuniary. This is what I had to shew wherein this order cannot conduce to that end, whereof it bears the intention.

I lastly proceed from the no good it can do, to the manifest hurt it causes, in being first the greatest discouragement and affront, that can be offer'd to learning and to learned men. It was the complaint and lamentation of

Prelats, upon every last breath of a motion to remove pluralities, and distribute more equally Church revennu's, that then all learning would be for ever dasht and discourag'd. But as for that opinion, I never found cause to think that the tenth part of learning stood or fell with the Clergy: nor could I ever but hold it for a sordid and unworthy speech of any Churchman who had a competency left him. If therefore ye be loath to dishearten utterly and discontent, not the mercenary crew of false pretenders to learning, but the free and ingenuous sort of such as evidently were born to study, and love lerning for it self, not for lucre, or any other end, but the service of God and of truth, and perhaps that lasting fame and perpetuity of praise which God and good men have consented shall be the reward of those whose publisht labours advance the good of mankind, then know, that so far to distrust the judgement & the honesty of one who hath but a common repute in learning, and never yet offended, as not to count him fit to print his mind without a tutor and examiner, lest he should drop a scism, or something of corruption, is the greatest displeasure and indignity to a free and knowing spirit that can be put upon him. What advantage is it to be a man over it is to be a boy at school if we have only scapt the ferular, to come under the fescu of an *Imprimatur?*[14] if serious and elaborat writings, as if they were no more then the theam of a Grammar lad under his Pedagogue must not be utter'd[15] without the cursory eyes of a temporizing and extemporizing licencer. He who is not trusted with his own actions, his drift not being known to evill, and standing to the hazard of law and penalty, has no great argument to think himself reputed in the Commonwealth wherin he was born, for other than a fool or a foreiner. When a man writes to the world, he summons up all his reason and deliberation to assist him; he searches, meditats, is industrious, and likely consults and conferrs with his judicious friends; after all which done he takes himself to be inform'd in what he writes, as well as any that writ before him; if in this the most consummat act of his fidelity and ripenesse, no years, no industry, no former proof of his abilities can bring him to that state of maturity, as not to be still mistrusted and suspected, unlesse he carry all his considerat diligence, all his midnight watchings, and expence of *Palladian*[16] oyl, to the hasty view of an unleasur'd licencer, perhaps much his younger, perhaps far his inferiour in judgement, perhaps one who never knew the labour of book-writing, and if he be not repulst, or slighted, must appear in Print like a punie[17] with his guardian, and his censors hand on the back of his title to be his bayl and surety, that he is no idiot, or seducer, it cannot be but a dishonor and derogation to the author, to the book, to the priviledge and dignity of Learning. And what if the author shall be one so copious of fancie, as to have many things well worth the adding, come into his mind after licencing, while the book is yet under the Presse, which not seldom happ'ns to the best and diligentest writers; and that perhaps a dozen times in one book. The Printer dares not go beyond his licenc't copy; so often then must the author trudge to his leav-giver, that those his new insertions may be viewd; and many a jaunt will be made, ere that licencer, for it must be the same man, can either be found, or found at leisure; mean while either the Presse must stand still, which is no small damage, or the author loose his accuratest thoughts, & send the book forth wors then he had made it, which to a diligent writer is the greatest melancholy and vexation that can befall. And how can a man teach with autority, which is the life of teaching, how can he be a Doctor in his book as he ought to be, or else had better be silent, whenas all he teaches, all he delivers, is but under the tuition, under the correction of his patriarchal licencer to blot or alter what precisely accords not with the hidebound humor which he calls his judgement. When every acute reader upon the first sight of a pedantick licence, will be ready with these like words to ding the book a coits distance from him, I hate a pupil teacher, I endure not an instructer that comes to me under the wardship of an overseeing fist. I know nothing of the licencer, but that I have his own hand here for his arrogance; who shall warrant me his judgement? The State Sir, replies the Stationer, but has a quick return, The State shall be my governours, but not my criticks; they may be mistak'n in the choice of a licencer, as easily as this licencer may be mistak'n in an author: This is some common stuffe; and he might adde from Sir *Francis Bacon*, That *such authoriz'd books are but the language of the times.*[18] For though a licencer should happ'n to be judicious more then ordnary, which will be a great jeopardy to the next succession, yet his very office, and his commission enjoyns him to let passe nothing but what is vulgarly receiv'd already. Nay, which is more lamentable, if the work of any deceased author, though never so famous in his life time, and even to this day, come to their hands for licence to be Printed, or Reprinted, if there be found in his book one sentence of a ventrous edge, utter'd in the height of zeal, and who knows whether it might not be the dictat of a divine Spirit, yet not suiting with every low decrepit humor of their own, though it were *Knox*[19] himself, the Reformer of a Kingdom that spake it, they will not pardon him their dash: the sense of that great man shall to all posterity be lost, for the fearfulnesse, or the presumptuous rashnesse of a perfunctory licencer. And to what an author this violence hath bin lately done, and in what book of greatest consequence to be faithfully publisht, I could now instance, but shall forbear till a more convenient season. Yet if these things be not resented seriously and timely by them who have the remedy in their power, but that such iron moulds as these shall have autority to knaw out the choisest periods of exquisitest books, and to commit such a treacherous fraud against the orphan remainders of worthiest men after death, the more sorrow will belong to that haples race of men, whose misfortune it is to have understanding. Henceforth let no man care to learn, or care to be more then worldly wise; for certainly in higher matters to be ignorant and slothfull, to a a a common stedfast dunce will be the only pleasant life, and only in request. . . .

Truth indeed came once into the world with her divine Master, and was a perfect shape most glorious to look on: but when he ascended, and his Apostles after him were laid asleep, then strait arose a wicked race of deceivers, who as that story goes of the *Ægyptian Typhon* with his conspirators, how they dealt with the good *Osiris*, took the virgin Truth, hewd her lovely form into a thousand peeces, and scatter'd them to the four winds. From that time ever since, the sad friends of Truth, such as durst appear, imitating the carefull search that *Isis* made for the mangl'd body of *Osiris*, went up and down gathering up limb by limb still as they could find them.[20] We have not yet found them all, Lords and Commons, nor ever shall doe, till her Masters second comming; he shall bring together every joynt and member, and shall mould them into an immortall feature of lovelines and perfection. Suffer not these licencing prohibitions to stand at every place of opportunity forbidding and disturbing them that continue seeking, that continue to do our obsequies to the torn body of our martyr'd Saint. We boast our light; but if we look not wisely on the Sun it self, it smites us into darknes. Who can discern those planets that are oft *Combust*,[21] and those stars of brightest magnitude that rise and set with the Sun, untill the opposite motion of their orbs bring them to such a place in the firmament, where they may be seen evning or morning. The light which we have gain'd, was giv'n us, not to be ever staring on, but by it to discover onward things more remote from our knowledge. It is not the unfrocking of a Priest, the unmitring of a Bishop, and the removing him from off the *Presbyterian* shoulders that will make us a happy Nation, no, if other things as great in the Church, and in the rule of life both economicall and politicall be not lookt into and reform'd, we have lookt so long upon the blaze that *Zuinglius* and *Calvin*[22] hath beacon'd up to us, that we are stark blind. There be who perpetually complain of schisms and sects, and make it such a calamity that any man dissents from their maxims. 'Tis their own pride and ignorance which causes the disturbing, who neither will hear with meeknes, nor can convince, yet all must be supprest which is not found in their *Syntagma*.[23] They are all troublers, they are the dividers of unity, who neglect and permit others to unite those dissever'd peeces which are yet wanting to the body of Truth. To be still searching what we know not, by what we know, still closing up truth to truth as we find it (for all her body is *homogeneal*,[24] and proportionall),[25] this is the golden rule in *Theology* as well as in Arithmetick, and makes up the best harmony in a Church; not the forc't and outward union of cold, and neutrall, and inwardly divided minds.

Sonnet XVII

When I consider how my light is spent,
  E're half my days, in this dark world and wide,
  And that one Talent[1] which is death to hide,
  Lodg'd with me useless, though my Soul more bent
To serve therewith my Maker, and present
  My true account, least he returning chide,
  Doth God exact day-labour, light deny'd,
  I fondly ask; But patience to prevent
That murmur, soon replies, God doth not need
  Either man's work or his own gifts, who best
  Bear his milde yoak, they serve him best, his State
Is Kingly. Thousands at his bidding speed
  And post o're Land and Ocean without rest:
  They also serve who only stand and waite.

John Milton (1608–74)

JOHN LOCKE

John Locke (1632–1704), Oxford don, physician, and civil servant, wrote on religious and political matters as well as philosophical. J. S. Mill called him 'the unquestioned founder of the analytic philosophy of mind.' Here in the opening chapters of Book III he investigates, with characteristic rigour, the nature of language and its relationship with experience and reality.

from *An Essay Concerning Human Understanding*

Of Words or Language in General

1. God having designed Man for a sociable Creature, made him not only with an inclination, and under a necessity to have fellowship with those of his own kind; but furnished him also with Language, which was to be the great Instrument, and common Tye of Society. *Man* therefore had by Nature his Organs so fashioned, as to be *fit to frame articulate Sounds*, which we call Words. But this was not enough to produce Language; for Parrots, and several other Birds, will be taught to make articulate Sounds distinct enough, which yet, by no means, are capable of Language.

2. Besides articulate Sounds therefore, it was farther necessary, that he should be *able to use these Sounds, as Signs of internal Conceptions*; and to make them stand as marks for the *Ideas* within his own Mind, whereby they might be made known to others, and the Thoughts of Men's Minds be conveyed from one to another.

3. But neither was this sufficient to make Words so useful as they ought to be. It is not enough for the perfection of Language, that Sounds can be made signs of *Ideas*, unless those *signs* can be so made use of, as *to comprehend several particular Things*: For the multiplication of Words would have perplexed their Use, had every particular thing need of a distinct name to be signified by. To remedy this inconvenience, Language had yet a farther improvement in the use of general Terms, whereby one word was made to mark a multitude of particular existences: Which advantageous use of Sounds was obtain'd only by the difference of the *Ideas* they were made signs of. Those names becoming general, which are made to stand for general *Ideas*, and those remaining particular, where the *Ideas* they are used for are particular.

4. Besides these Names which stand for *Ideas*, there be other words which Men make use of, not to signify any *Idea*, but the want or absence of some

Ideas simple or complex, or all *Ideas* together; such as are *Nihil* in Latin, and in English, *Ignorance* and *Barrenness*. All which negative or privative Words, cannot be said properly to belong to, or signify no *Ideas*: for then they would be perfectly insignificant Sounds; but they relate to positive *Ideas*, and signify their absence.

5. It may also lead us a little towards the Original of all our Notions and Knowledge, if we remark, how great a dependance our *Words* have on common sensible *Ideas*; and how those, which are made use of to stand for Actions and Notions quite removed from sense, *have their rise from thence, and from obvious sensible* Ideas *are transferred to more abstruse significations*, and made to stand for *Ideas* that come not under the cognizance of our senses; *v.g.*[1] to *Imagine, Apprehend, Comprehend, Adhere, Conceive, Instill, Disgust, Disturbance, Tranquillity*, etc. are all Words taken from the Operations of sensible Things, and applied to certain Modes of Thinking. *Spirit*, in its primary signification, is Breath; *Angel*, a Messenger: And I doubt not, but if we could trace them to their sources, we should find, in all Languages, the names, which stand for Things that fall not under our Senses, to have had their first rise from sensible *Ideas*. By which we may give some kind of guess, what kind of Notions they were, and whence derived, which filled their Minds, who were the first Beginners of Languages; and now Nature, even in the naming of Things, unawares suggested to Men the Originals and Principles of all their Knowledge: whilst, to give Names, that might make known to others any Operations they felt in themselves, or any other *Ideas*, that came not under their Senses, they were fain to borrow Words from ordinary known *Ideas* of Sensation, by that means to make others the more easily to conceive those Operations they experimented in themselves, which made no outward sensible appearances; and then when they had got known and agreed Names, to signify those internal Operations of their own Minds, they were sufficiently furnished to make known by Words, all their other *Ideas*; since they could consist of nothing, but either of outward sensible Perceptions, or of the inward Operations of their Minds about them; we having, as has been proved, no *Ideas* at all, but what originally come either from sensible Objects without, or what we feel within our selves, from the inward Workings of our own Spirits, which we are conscious to our selves of within.

6. But to understand better the use and force of Language, as subservient to Instruction and Knowledge, it will be convenient to consider,

First, To what it is that Names, in the use of Language, are immediately applied.

Secondly, Since all (except proper) Names are general, and so stand not for this or that single thing; but for sorts and ranks of Things, it will be necessary to consider, in the next place, what the Sorts or Kinds, or, if you rather like the Latin Names, *what the Species and Genera of Things* are;

wherein they consist; and how they come to be made. These being (as they ought) well looked into, we shall the better come to find the right use of Words; the natural Advantages and Defects of Language; and the remedies that ought to be used, to avoid the inconveniences of obscurity or uncertainty in the signification of Words, without which, it is impossible to discourse with any clearness, or order, concerning Knowledge: Which being conversant about Propositions, and those most commonly universal ones, has greater connexion with Words, than perhaps is suspected.

These Considerations therefore, shall be the matter of the following Chapters.

Of the Signification of Words

1. Man, though he have great variety of Thoughts, and such, from which others, as well as himself, might receive Profit and Delight; yet they are all within his own Breast, invisible, and hidden from others, nor can of themselves be made appear. The Comfort, and Advantage of Society, not being to be had without Communication of Thoughts, it was necessary, that Man should find out some external sensible Signs, whereby those invisible *Ideas*, which his thoughts are made up of, might be made known to others. For this purpose, nothing was so fit, either for Plenty or Quickness, as those articulate Sounds, which with so much Ease and Variety, he found himself able to make. Thus we may conceive how *Words*, which were by Nature so well adapted to that purpose, come to be made use of by Men, as *the Signs of* their *Ideas*, not by any natural connexion, that there is between particular articulate Sounds and certain *Ideas*, for then there would be but one Language among all Men; but by a voluntary imposition, whereby such a word is made arbitrarily the Mark of such an *Idea*. The use then of Words, is to be sensible Marks of *Ideas*; and the *Ideas* they stand for, are their proper and immediate Signification.

2. The use Men have of these Marks, being either to record their own Thoughts for the Assistance of their own Memory; or as it were, to bring out their *Ideas*, and lay them before the view of others: *Words in their primary or immediate Signification, stand for nothing, but the* Ideas *in the Mind of him that uses them*, how imperfectly soever, or carelessly those *Ideas* are collected from the Things, which they are supposed to represent. When a Man speaks to another, it is, that he may be understood; and the end of Speech is, that those sounds, as Marks, may make known his *Ideas* to the Hearer. That then which Words are the Marks of, are the *Ideas* of the Speaker: Nor can any one apply them, as Marks, immediately to any thing else, but the *Ideas*, that he himself hath: for this would be to make them Signs of his own Conceptions, and yet apply them to other *Ideas* at the same time;

and so in effect, to have no Signification at all. Words being voluntary Signs, they cannot be voluntary Signs imposed by him on Things he knows not. That would be to make them Signs of nothing, Sounds without Signification. A Man cannot make his Words the Signs either of Qualities in Things, or of Conceptions in the Mind of another, whereof he has none in his own. Till he has some *Ideas* of his own, he cannot suppose them to correspond with the Conceptions of another Man; nor can he use any Signs for them: For thus they would be the Signs of he knows not what, which is in Truth to be the Signs of nothing. But when he represents to himself other Men's *Ideas*, by some of his own, if he consent to give them the same Names, that other Men do, 'tis still to his own *Ideas*, to *Ideas* that he has, and not to *Ideas* that he has not.

3. This is so necessary in the use of Language, that in this respect, the Knowing, and the Ignorant; the Learned, and Unlearned, use the *Words* they speak (with any meaning) all alike. They, *in every Man's Mouth, stand for the* Ideas *he has*, and which he would express by them. A Child having taken notice of nothing in the Metal he hears called Gold, but the bright shining yellow colour, he applies the Word Gold only to his own *Idea* of that Colour, and nothing else; and therefore calls the same Colour in a Peacocks Tail, Gold. Another that hath better observed, adds to shining yellow, great Weight; And then the Sound Gold, when he uses it, stands for a complex *Idea* of a shining Yellow and very weighty Substance. Another adds to those Qualities, Fusibility: and then the Word Gold to him signifies a Body, bright, yellow, fusible, and very heavy. Another adds Malleability. Each of these uses equally the Word Gold, when they have Occasion to express the *Idea*, which they have apply'd it to: But it is evident, that each can apply it only to his own *Idea*; nor can he make it stand, as a Sign of such a complex *Idea*, as he has not.

4. But though Words, as they are used by Men, can properly and immediately signify nothing but the *Ideas*, that are in the Mind of the Speaker; yet they in their Thoughts give them a secret reference to two other things.

First, they suppose their Words to be Marks of the Ideas *in the Minds also of other Men, with whom they communicate*: For else they should talk in vain, and could not be understood, if the Sounds they applied to one *Idea*, were such, as by the Hearer, were applied to another, which is to speak two Languages. But in this, Men stand not usually to examine, whether the *Idea* they, and those they discourse with have in their Minds, be the same: But think it enough, that they use the Word, as they imagine, in the common Acceptation of that Language; in which case they suppose, that the *Idea*, they make it a Sign of, is precisely the same, to which the Understanding Men of that Country apply that Name.

5. *Secondly*, Because *Men* would not be thought to talk *barely* of their

own Imaginations, but of Things as really they are; therefore they *often suppose their Words to stand also for the reality of Things*. But this relating more particularly to Substances, and their Names, as perhaps the former does to simple *Ideas* and Modes, we shall speak of these two different ways of applying Words more at large, when we come to treat of the Names of mixed Modes, and Substances, in particular: Though give me leave here to say, that it is a perverting the use of Words, and brings unavoidable Obscurity and Confusion into their Signification, whenever we make them stand for any thing, but those *Ideas* we have in our own Minds.

6. Concerning Words also it is farther to be considered. *First*, That they being immediately the Signs of Mens *Ideas*; and by that means, the Instruments whereby Men communicate their Conceptions, and express to one another those Thoughts and Imaginations, they have within their own Breasts, *there comes by constant use*, to be such *a Connexion between certain Sounds, and the* Ideas *they stand for*, that the Names heard, almost as readily excite certain *Ideas*, as if the Objects themselves, which are apt to produce them, did actually affect the Senses. Which is manifestly so in all obvious sensible Qualities; and in all Substances, that frequently, and familiarly occur to us.

7. *Secondly*, That though the proper and immediate Signification of Words, are *Ideas* in the mind of the Speaker; yet because by familiar use from our Cradles, we come to learn certain articulate Sounds very perfectly, and have them readily on our Tongues, and always at hand in our Memories; but yet are not always careful to examine, or settle their Significations perfectly, it *often* happens that *Men*, even when they would apply themselves to an attentive Consideration, do *set their Thoughts more on Words than Things*. Nay, because Words are many of them learn'd, before the *Ideas* are known for which they stand: Therefore some, not only Children, but Men, speak several Words, no otherwise than Parrots do, only because they have learn'd them, and have been accustomed to those Sounds. But so far as Words are of Use and Signification, so far is there a constant connexion between the Sound and the *Idea*; and a Designation, that the one stand for the other: without which Application of them, they are nothing but so much insignificant Noise.

8. *Words* by long and familiar use, as has been said, come to excite in Men certain *Ideas*, so constantly and readily, that they are apt to suppose a natural connexion between them. But that they *signify* only Men's peculiar *Ideas*, and that *by a perfectly arbitrary Imposition*, is evident, in that they often fail to excite in others (even that use the same Language) the same *Ideas*, we take them to be the Signs of: And every Man has so inviolable a Liberty, to make Words stand for what *Ideas* he pleases, that no one hath the Power to make others have the same *Ideas* in their Minds, that he has, when they use the same Words, that he does. And therefore the great *Augustus* himself, in the

Possession of that Power which ruled the World, acknowledged, he could not make a new Latin Word: which was as much as to say, that he could not arbitrarily appoint, what *Idea* any Sound should be a Sign of, in the Mouths and common Language of his Subjects. 'Tis true, common use, by a tacit Consent, appropriates certain Sounds to certain *Ideas* in all Languages, which so far limits the signification of that Sound, that unless a Man applies it to the same *Idea*, he does not speak properly: And let me add, that unless a Man's Words excite the same *Ideas* in the Hearer, which he makes them stand for in speaking, he does not speak intelligibly. But whatever be the consequence of any Man's using of Words differently, either from their general Meaning, or the particular Sense of the Person to whom he addresses them, this is certain, their signification, in his use of them, is limited to his *Ideas*, and they can be Signs of nothing else.

To the Memory of Mr Oldham

Farewell, too little and too lately known,
Whom I began to think and call my own:
For sure our Souls were near alli'd, and thine
Cast in the same poetick mold with mine.
One common Note on either Lyre did strike,
And Knaves and Fools we both abhorr'd alike.
To the same Goal did both our Studies drive:
The last set out the soonest did arrive.
Thus *Nisus* fell upon the slippery place,
Whilst his young Friend perform'd and won the Race.[1]
O early ripe! to thy abundant Store
What could advancing Age have added more?
It might (what Nature never gives the Young)
Have taught the Numbers[2] of thy Native Tongue.
But Satire needs not those, and Wit will shine
Through the harsh Cadence of a rugged Line.
A noble Error, and but seldom made,
When Poets are by too much force betray'd.
Thy gen'rous Fruits, though gather'd ere their prime,
Still shew'd a Quickness; and maturing Time
But mellows what we write to the dull Sweets of Rhyme.
Once more, hail, and farewell! farewell, thou young,
But ah! too short, *Marcellus*[3] of our Tongue!
Thy Brows with Ivy and with Laurels bound;
But Fate and gloomy Night encompass thee around.

John Dryden (1631–1700)

JONATHAN SWIFT

Jonathan Swift (1667–1745), widely considered the greatest of English prose writers, virtually ruled England by the power of his pen during the final years of Queen Anne's reign. His chief weapon of irony found its most famous expression in *Gulliver's Travels*. The following contribution to a fashionable literary periodical embodies one of several unsuccessful attempts he made to legislate for the purification of the English language: '. . . he published a *Proposal for correcting, improving, and ascertaining the English Tongue*, in a Letter to the Earl of Oxford; written without much knowledge of the general nature of Language, and without any accurate enquiry into the history of other tongues. The certainty and stability which, contrary to all experience, he thinks attainable, he proposes to secure by instituting an Academy; the decrees of which every man would have been willing, and many would have been proud to disobey, and which, being renewed by successive elections, would in a short time have differed from itself.' (Dr Johnson, *Life of Swift*, 1781). Only a little of the slang Swift attacks here has survived. This piece on the corruption of the English language was first printed in *The Tatler*, Number CCXXX, Thursday September 28th, 1710.

from *The Tatler*

From my own Apartment, September 27.

The following Letter has laid before me many great and manifest Evils in the World of Letters which I had overlooked; but they open to me a very busie Scene, and it will require no small Care and Application to amend Errors which are become so universal. The Affectation of Politeness is exposed in this Epistle with a great deal of Wit and Discernment; so that whatever Discourses I may fall into hereafter upon the Subjects the Writer treats of, I shall at present lay the Matter before the World without the least Alteration from the Words of my Correspondent.

To Isaac Bickerstaff [1] *Esq;*
SIR,

There are some Abuses among us of great Consequence, the Reformation of which is properly your Province, tho', as far as I have been conversant in your Papers, you have not yet considered them. These are, the deplorable Ignorance that for some Years hath reigned among our English Writers, the great Depravity of our Taste, and the continual Corruption of our Style. I say nothing here of those who handle particular Sciences, Divinity, Law, Phisick, and the like; I mean, the Traders in History and Politicks, and the Belles Lettres; together with those by whom Books are not translated, but (as the common Expressions are) Done out of French, Latin, or other

Language, and Made English. I cannot but observe to you, That till of late Years a Grub-street[2] Book was always bound in Sheep-skin, with suitable Print and Paper, the Price never above a Shilling, and taken off wholly by common Tradesmen, or Country Pedlars. But now they appear in all Sizes and Shapes, and in all Places. They are handed about from Lapfulls in every Coffee-house to Persons of Quality, are shewn in Westminster-Hall and the Court of Requests. You may see them gilt, and in Royal Paper, of Five or Six hundred Pages, and rated accordingly. I would engage to furnish you with a Catalogue of English Books published within the Compass of Seven Years past, which at the first Hand would cost you a Hundred Pounds, wherein you shall not be able to find Ten Lines together of common Grammar or common Sense.

These Two Evils, Ignorance and Want of Taste, have produced a Third; I mean, the continual Corruption of our English Tongue, which, without some timely Remedy, will suffer more by the false Refinements of Twenty Years past, than it hath been improved in the foregoing Hundred: And this is what I design chiefly to enlarge upon, leaving the former Evils to your Animadversion.

But instead of giving you a List of the late Refinements crept into our Language, I here send you the Copy of a Letter I received some Time ago from a most accomplished Person in this Way of Writing, upon which I shall make some Remarks. It is in these Terms.[3]

SIR,

'I cou'dn't get the Things you sent for all *about Town*. . . . I *thôt* to *ha'* come down my self, and then *I'd ha' brôut 'um*; but I *han't don't*, and I believe I *can't do't*, that's *Pozz*. . . . *Tom* begins to *gi'mself Airs* because *he's* going with the *Plenipo's*. . . . 'Tis said, the *French* King will *bamboozl' us agen*, which *causes many Speculations*. The *Jacks*, and others of that *Kidney*, are very *uppish*, and *alert upon't*, as you may see by their *Phizz's*. . . . *Will Hazzard* has got the *Hipps*, having lost *to the Tune of* Five hundr'd Pound, *thô* he understands Play very well, *no body better*. He has promis't me upon *Rep*, to leave off Play; but you know 'tis a Weakness *he's* too apt to *give into*, *thô* he has as much Wit as any Man, *no body more*. He has lain *incog* ever since. . . . The *Mobb's* very quiet with us now. . . . I believe you *thot* I *banter'd* you in my Last like a *Country Put*. . . . I *sha'n't* leave Town this Month, *&c*.

This Letter is in every Point an admirable Pattern of the present polite Way of Writing; nor is it of less Authority for being an Epistle. You may gather every Flower in it, with a Thousand more of equal Sweetness, from the Books, Pamphlets, and single Papers, offered us every Day in the Coffee-houses: And these are the Beauties introduced to supply the Want of Wit, Sense, Humour and Learning, which formerly were looked upon as

Qualifications for a Writer. If a Man of Wit, who died Forty Years ago, were to rise from the Grave on Purpose, How would he be able to read this Letter? And after he had got through that Difficulty, How would he be able to understand it? The first Thing that strikes your Eye is the Breaks at the End of almost every Sentence; of which I know not the Use, only that it is a Refinement, and very frequently practised. Then you will observe the Abbreviations and Elisions, by which Consonants of most obdurate Sound are joined together, without one softening Vowel to intervene; and all this only to make one Syllable of two, directly contrary to the Example of the Greeks and Romans; altogether of the Gothick Strain, and a natural Tendency towards relapsing into Barbarity, which delights in Monosyllables, and uniting of Mute Consonants; as it is observable in all the Northern Languages. And this is still more visible in the next Refinement, which consists in pronouncing the first Syllable in a Word that has many, and dismissing the rest; such as *Phizz, Hipps, Mobb, Poz. Rep.*[4] and many more; when we are already overloaded with Monosyllables, which are the Disgrace of our Language. Thus we cram one Syllable, and cut off the rest; as the Owl fatten'd her Mice, after she had bit off their Legs to prevent them running away; and if ours be the same Reason for maiming our Words, it will certainly answer the End; for I am sure no other Nation will desire to borrow them. Some Words are hitherto but fairly split, and therefore only in their Way to Perfection, as *Incog* and *Plenipo*: But in a short Time 'tis to be hoped they will be further dock'd to *Inc* and *Plen*. This Reflexion had made me of late Years very impatient for a Peace, which I believe would save the Lives of many brave Words, as well as Men. The War has introduced Abundance of Polysyllables, which will never be able to live many more Campagnes; *Speculations, Operations, Preliminaries, Ambassadors, Pallisadoes, Communication, Circumvallation, Battalions*, as numerous as they are, if they attack us too frequently in our Coffee-houses, we shall certainly put them to Flight, and cut off the Rear.

The Third Refinement observable in the Letter I send you, consists in the Choice of certain Words invented by some Pretty Fellows; such as *Banter, Bamboozle, Country Put*, and *Kidney*, as it is there applied; some of which are now struggling for the Vogue, and others are in Possession of it. I have done my utmost for some Years past to stop the Progress of *Mobb* and *Banter*, but have been plainly borne down by Numbers, and betrayed by those who promised to assist me.

In the last Place, you are to take Notice of certain choice Phrases scattered through the Letter; some of them tolerable enough, till they were worn to Rags by servile Imitators. You might easily find them, though they were not in a different Print, and therefore I need not disturb them.

These are the false Refinements in our Style which you ought to correct: First, by Argument and fair Means; but if those fail, I think you are to make

Use of your Authority as Censor, and by an annual *Index Expurgatorius*[5] expunge all Words and Phrases that are offensive to good Sense, and condemn those barbarous Mutilations of Vowels and Syllables. In this last point the usual Pretence is, that they spell as they speak; A noble Standard for Language! to depend upon the Caprice of every Coxcomb, who, because Words are the Cloathing of our Thoughts, cuts them out, and shapes them as he pleases, and changes them oftner than his Dress. I believe, all reasonable People would be content that such Refiners were more sparing in their Words, and liberal in their Syllables: And upon this Head I should be glad you would bestow some Advice upon several young Readers in our Churches, who coming up from the University, full fraught with Admiration of our Town Politeness, will needs correct the Style of their Prayer Books. In reading the Absolution, they are very careful to say *pardons* and *absolves*; and in the Prayer for the Royal Family, it must be, *endue 'um, enrich 'um, prosper 'um*, and *bring 'um*. Then in their Sermons they use all the modern Terms of Art, *Sham, Banter, Mob, Bubble, Bully, Cutting, Shuffling*, and *Palming*, all which, and many more of the like Stamp, as I have heard them often in the Pulpit from such young Sophisters, so I have read them in some of those Sermons that have made most Noise of late. The Design, it seems, is to avoid the dreadful Imputation of Pedantry, to shew us, that they know the Town, understand Men and Manners, and have not been poring upon old unfashionable Books in the University.

I should be glad to see you the Instrument of introducing into our Style that Simplicity which is the best and truest Ornament of most Things in Life, which the politer Ages always aimed at in their Building and Dress, (*Simplex munditiis*)[6] as well as their Productions of Wit. 'Tis manifest, that all new, affected Modes of Speech, whether borrowed from the Court, the Town, or the Theatre, are the first perishing Parts in any Language, and, as I could prove by many Hundred Instances, have been so in ours. The Writings of Hooker, who was a Country Clergymen, and of Parsons the Jesuit, both in the Reign of Queen Elizabeth, are in a Style that, with very few Allowances, would not offend any present Reader; much more clear and intelligible than those of Sir H. Wotton, Sir Robert Naunton, Osborn, Daniel the Historian, and several others who writ later; but being Men of the Court, and affecting the Phrases then in Fashion, they are often either not to be understood, or appear perfectly ridiculous.

What Remedies are to be applied to these Evils I have not Room to consider, having, I fear, already taken up most of your Paper. Besides, I think it is our Office only to represent Abuses, and yours to redress them. I am, with great Respect,

SIR,

Your, &c.

from *On the Death of Dr Swift*

Suppose me dead; and then suppose
A club assembled at the Rose;
Where, from discourse of this and that,
I grow the subject of their chat.
And while they toss my name about,
With favour some, and some without,
One, quite indifferent in the cause,
My character impartial draws:
   'The Dean, if we believe report,
Was never ill receiv'd at court.
As for his works in verse and prose
I own myself no judge of those;
Nor can I tell what critics thought 'em;
But this I know, all people bought 'em
As with a moral view design'd
To cure the vices of mankind.
And, if he often miss'd his aim,
The world must own it, to their shame,
The praise is his, and theirs the blame.' —
'Sir, I have heard another story;
He was a most confounded Tory,
And grew, or he is much belied,
Extremely dull, before he died.' —
   'Can we the Drapier then forget?
Is not our nation in his debt?
'Twas he that writ the Drapier's letters!¹ —
   'He should have left them for his betters,
We had a hundred abler men,
Nor need depend upon his pen.' —
'Say what you will about his reading,
You never can defend his breeding;
Who in his satires running riot,
Could never leave the world in quiet;
Attacking, when he took the whim,
Court, city, camp — all one to him.' —
   'But why should he, except he slobber't,
Offend our patriot, great Sir Robert,
Whose counsels aid the sov'reign power
To save the nation every hour?
What scenes of evil he unravels

In satires, libels, lying travels!
Not sparing his own clergy-cloth,
But eats into it, like a moth!' —
    'His vein, ironically grave,
Exposed the fool, and lash'd the knave.
To steal a hint was never known,
But what he writ was all his own.'

Jonathan Swift (1667–1745)

SAMUEL JOHNSON

Samuel Johnson (1709–84), the leading literary scholar and critic of his time, published his great *Dictionary of the English Language* in 1755. In its *Preface*, in his critical biographies entitled *The Lives of the Poets*, in his influential Shakespearian criticism, and in his prolific output of occasional writings he works out a profoundly pondered and intelligently balanced response to the problems and paradoxes of language and writing.

from *The Preface to the Dictionary*

It is the fate of those who toil at the lower employments of life to be rather driven by the fear of evil than attracted by the prospect of good; to be exposed to censure, without hope of praise; to be disgraced by miscarriage, or punished for neglect, where success would have been without applause, and diligence without reward.

Among these unhappy mortals is the writer of dictionaries; whom mankind have considered, not as the pupil, but the slave of science, the pioneer of literature, doomed only to remove rubbish and clear obstructions from the paths through which Learning and Genius press forward to conquest and glory, without bestowing a smile on the humble drudge that facilitates their progress. Every other author may aspire to praise; the lexicographer can only hope to escape reproach, and even this negative recompense has been yet granted to very few.

I have, notwithstanding this discouragement, attempted a dictionary of the English language, which, while it was employed in the cultivation of every species of literature, has itself been hitherto neglected; suffered to spread, under the direction of chance, into wild exuberance, resigned to the tyranny of time and fashion, and exposed to the corruptions of ignorance, and caprices of innovation.

When I took the first survey of my undertaking, I found our speech copious without order, and energetic without rules: wherever I turned my view, there was perplexity to be disentangled, and confusion to be regulated; choice was to be made out of boundless variety, without any established principle of selection; adulterations were to be detected, without a settled test of purity, and modes of expression to be rejected or received, without the suffrages of any writers of classical reputation or acknowledged authority.

Having therefore no assistance but from general grammar, I applied myself to the perusal of our writers; and noting whatever might be of use to ascertain or illustrate any word or phrase, accumulated in time the materials of a dictionary, which, by degrees, I reduced to method, establishing to myself, in the progress of the work, such rules as experience and analogy suggested to me; experience, which practice and observation were continually increasing; and analogy, which, though in some words obscure, was evident in others.

In adjusting the ORTHOGRAPHY, which has been to this time unsettled and fortuitous, I found it necessary to distinguish those irregularities that are inherent in our tongue, and perhaps coeval with it, from others which ignorance or negligence of later writers has produced. Every language has its anomalies, which, though inconvenient, and in themselves once unnecessary, must be tolerated among the imperfections of human things, and which require only to be registered, that they may not be increased, and ascertained, that they may not be confounded: but every language has likewise its improprieties and absurdities, which it is the duty of the lexicographer to correct or proscribe.

As language was at its beginning merely oral, all words of necessary or common use were spoken before they were written; and while they were unfixed by any visible signs, must have been spoken with great diversity, as we now observe those who cannot read to catch sounds imperfectly, and utter them negligently. When this wild and barbarous jargon was first reduced to an alphabet, every penman endeavoured to express, as he could, the sounds which he was accustomed to pronounce or to receive, and vitiated in writing such words as were already vitiated in speech. The powers of the letters, when they were applied to a new language, must have been vague and unsettled, and, therefore, different hands would exhibit the same sound by different combinations.

From this uncertain pronunciation arise, in a great part, the various dialects of the same country, which will always be observed to grow fewer and less different, as books are multiplied; and from this arbitrary representation of sounds by letters proceeds that diversity of spelling, observable in the Saxon remains, and, I suppose, in the first books of every nation, which perplexes or destroys analogy, and produces anomalous formations that being once incorporated can never be afterward dismissed or reformed. . . .

In this part of the work, where caprice has long wantoned without control, and vanity sought praise by petty reformation, I have endeavoured to proceed with a scholar's reverence for antiquity, and a grammarian's regard to the genius of our tongue. I have attempted few alterations, and among those few, perhaps, the greater part is from the modern to the ancient practice; and I hope I may be allowed to recommend to those whose thoughts have been, perhaps, employed too anxiously on verbal singularities, not to disturb, upon narrow views, or for minute propriety, the orthography of their fathers. It has been asserted that for the law to be *known* is of more importance than to be *right*. 'Change,' says Hooker, 'is not made without inconvenience, even from worse to better.'[1] There is in constancy and stability a general and lasting advantage, which will always overbalance the slow improvements of gradual correction. Much less ought our written language to comply with the corruptions of oral utterance, or copy that which every variation of time or place makes different from itself, and imitate those changes which will again be changed, while imitation is employed in observing them.

This recommendation of steadiness and uniformity does not proceed from an opinion that particular combinations of letters have much influence on human happiness; or that truth may not be successfully taught by modes of spelling fanciful and erroneous: I am not yet so lost in lexicography as to forget that *words are the daughters of earth, and that things are the sons of heaven*. Language is only the instrument of science, and words are but the signs of ideas: I wish, however, that the instrument might be less apt to decay, and that signs might be permanent, like the things which they denote. . . .

The solution of all difficulties, and the supply of all defects, must be sought in the examples subjoined to the various senses of each word, and ranged according to the time of their authors.

When first I collected these authorities, I was desirous that every quotation should be useful to some other end than the illustration of a word; I therefore extracted from philosophers principles of science; from historians remarkable facts; from chemists complete processes; from divines striking exhortations; and from poets beautiful descriptions. Such is design, while it is yet at a distance from execution. When the time called upon me to range this accumulation of elegance and wisdom into an alphabetical series, I soon discovered that the bulk of my volumes would fright away the student, and was forced to depart from my scheme of including all that was pleasing or useful in English literature, and reduce my transcripts very often to clusters of words in which scarcely any meaning is retained: thus to the weariness of copying, I was condemned to add the vexation of expunging. Some passages I have yet spared which may relieve the labour of verbal searches, and intersperse with verdure and flowers the dusty deserts of barren philology.

The examples, thus mutilated, are no longer to be considered as conveying the sentiments or doctrine of their authors; the word for the sake of which they are inserted, with all its appendant clauses, has been carefully preserved; but it may sometimes happen, by hasty detruncation, that the general tendency of the sentence may be changed: the divine may desert his tenets, or the philosopher his system.

Some of the examples have been taken from writers who were never mentioned as masters of elegance, or models of style; but words must be sought where they are used; and in what pages eminent for purity can terms of manufacture or agriculture be found? Many quotations serve no other purpose than that of proving the bare existence of words, and are therefore selected with less scrupulousness than those which are to teach their structures and relations.

My purpose was to admit no testimony of living authors, that I might not be misled by partiality, and that none of my contemporaries might have reason to complain; nor have I departed from this resolution but when some performance of uncommon excellence excited my veneration, when my memory supplied me from late books with an example that was wanting, or when my heart, in the tenderness of friendship, solicited admission for a favourite name.

So far have I been from any care to grace my pages with modern decorations that I have studiously endeavoured to collect examples and authorities from the writers before the Restoration, whose works I regard as *the wells of English undefiled*,[2] as the pure sources of genuine diction. Our language, for almost a century, has, by the concurrence of many causes, been gradually departing from its original Teutonic character, and deviating towards a Gallic structure and phraseology, from which it ought to be our endeavour to recall it, by making our ancient volumes the ground-work of style, admitting among the additions of later times only such as may supply real deficiencies, such as are readily adopted by the genius of our tongue, and incorporate easily with our native idioms.

But as every language has a time of rudeness antecedent to perfection, as well as of false refinement and declension, I have been cautious lest my zeal for antiquity might drive me into times too remote, and crowd my book with words now no longer understood. I have fixed Sidney's work for the boundary beyond which I make few excursions. From the authors which rose in the time of Elizabeth, a speech might be formed adequate to all the purposes of use and elegance. If the language of theology were extracted from Hooker and the translation of the Bible; the terms of natural knowledge from Bacon; the phrases of policy, war, and navigation from Raleigh; the dialect of poetry and fiction from Spenser and Sidney; and the diction of common life from Shakespeare, few ideas would be lost to mankind for want of English words in which they might be expressed. . . .

Total and sudden transformations of a language seldom happen; conquests and migrations are now very rare; but there are other causes of change, which, though slow in their operation, and invisible in their progress, are perhaps as much superior to human resistance as the revolutions of the sky, or intumescence of the tide. Commerce, however necessary, however lucrative, as it depraves the manners, corrupts the language; they that have frequent intercourse with strangers, to whom they endeavour to accommodate themselves, must in time learn a mingled dialect, like the jargon which serves the traffickers on the Mediterranean and Indian coasts. This will not always be confined to the exchange, the warehouse, or the port, but will be communicated by degrees to other ranks of the people, and be at last incorporated with the current speech.

There are likewise internal causes equally forcible. The language most likely to continue long without alteration would be that of a nation raised a little, and but a little, above barbarity, secluded from strangers, and totally employed in procuring the conveniences of life; either without books, or, like some of the Mahometan countries, with very few: men thus busied and unlearned, having only such words as common use requires, would perhaps long continue to express the same notions by the same signs. But no such constancy can be expected in a people polished by arts, and classed by subordination, where one part of the community is sustained and accommodated by the labour of the other. Those who have much leisure to think will always be enlarging the stock of ideas; and every increase of knowledge, whether real or fancied, will produce new words, or combinations of words. When the mind is unchained from necessity, it will range after convenience; when it is left at large in the fields of speculation, it will shift opinions; as any custom is disused, the words that expressed it must perish with it; as any opinion grows popular, it will innovate speech in the same proportion as it alters practice.

As by the cultivation of various sciences a language is amplified, it will be more furnished with words deflected from their original sense; the geometrician will talk of a courtier's zenith, or the eccentric virtue of a wild hero, and the physician of sanguine expectations and phlegmatic delays. Copiousness of speech will give opportunities to capricious choice, by which some words will be preferred, and others degraded; vicissitudes of fashion will enforce the use of new or extend the signification of known terms. The tropes of poetry will make hourly encroachments, and the metaphorical will become the current sense: pronunciation will be varied by levity or ignorance, and the pen must at length comply with the tongue; illiterate writers will at one time or other, by public infatuation, rise into renown, who, not knowing the original import of words, will use them with colloquial licentiousness, confound distinction, and forget propriety. As politeness increases, some expressions will be considered as too gross and vulgar for the delicate, others as too formal and ceremonious for the gay and airy; new phases are therefore adopted which must, for the same reasons, be in time dismissed. Swift, in his petty treatise[3] on the English language, allows that new words must sometimes be introduced, but proposes than none should be suffered to become obsolete. But what makes a word obsolete, more than general agreement to forbear it? and how shall it be continued, when it conveys an offensive idea, or recalled again into the mouths of mankind, when it has once become unfamiliar by disuse, and unpleasing by unfamiliarity?

There is another cause of alteration more prevalent than any other, which yet in the present state of the world cannot be obviated. A mixture of two languages will produce a third distinct from both; and they will always be mixed, where the chief part of education, and the most conspicuous accomplishment, is skill in ancient or in foreign tongues. He that has long cultivated another language will find its words and combinations crowd upon his memory; and hate or negligence, refinement or affectation, will obtrude borrowed terms and exotic expressions.

The greatest pest of speech is frequency of translation. No book was ever turned from one language into another without imparting something of its native idiom; this is the most mischievous and comprehensive innovation; single words may enter by thousands, and the fabric of the tongue continue the same, but new phraseology changes much at once; it alters not the single stones of the building, but the order of the columns. If an academy should be established for the cultivation of our style, which I, who can never wish to see dependence multiplied, hope the spirit of English liberty will hinder or destroy, let them, instead of compiling grammars and dictionaries, endeavour, with all their influence, to stop the licence of translators, whose idleness and ignorance, if it be suffered to proceed, will reduce us to babble a dialect of France.

If the changes we fear be thus irresistible, what remains but to acquiesce with silence, as in the other insurmountable distresses of humanity? It remains that we retard what we cannot repel, that we palliate what we cannot cure. Life may be lengthened by care, though death cannot be ultimately defeated: tongues, like governments, have a natural tendency to degeneration; we have long preserved our constitution, let us make some struggles for our language.

In hope of giving longevity to that which its own nature forbids to be immortal, I have devoted this book, the labour of years, to the honour of my country, that we may no longer yield the palm of philology without a contest to the nations of the continent. The chief glory of every people arises from its authors: whether I shall add anything by my own writings to the reputation of English literature must be left to time: much of my life has been lost under the pressure of disease; much has been trifled away; and much has always been spent in provision for the day that was passing over me; but I shall not think my employment useless or ignoble, if by my assistance foreign nations, and distant ages, gain access to the propagators of knowledge, and understand the teachers of truth; if my labours afford light to the repositories of science, and add celebrity to Bacon, to Hooker, to Milton, and to Boyle.

When I am animated by this wish, I look with pleasure on my book, however defective, and deliver it to the world with the spirit of a man that has endeavoured well. That it will immediately become popular I have not promised to myself: a few wild blunders, and risible absurdities, from which no work of such multiplicity was ever free, may for a time furnish folly with laughter, and harden ignorance in contempt; but useful diligence will at last prevail, and there never can be wanting some who distinguish desert; who will consider that no dictionary of a living tongue ever can be perfect, since while it is hastening to publication, some words are budding, and some falling away; that a whole life cannot be spent upon syntax and etymology, and that even a whole life would not be sufficient; that he whose design includes whatever language can express must often speak of what he does not understand; that a writer will sometimes be hurried by eagerness to the end, and sometimes faint with weariness under a task which Scaliger compares to the labours of the anvil and the mine; that what is obvious is not always known, and what is known is not always present; that sudden fits of inadvertency will surprise vigilance, slight avocations will seduce attention, and casual eclipses of the mind will darken learning; and that the writer shall often in vain trace his memory at the moment of need for that which yesterday he knew with intuitive readiness, and which will come uncalled into his thoughts tomorrow.

In this work, when it shall be found that much is omitted, let it not be forgotten that much likewise is performed; and though no book was ever spared out of tenderness to the author, and the world is little solicitous to know whence proceeded the faults of that which it condemns; yet it may gratify curiosity to inform it that the *English Dictionary* was written with little assistance of the learned, and without any patronage of the great; not in the soft obscurities of retirement, or under the shelter of academic bowers, but amidst inconvenience and distraction, in sickness and in sorrow: and it may repress the triumph of malignant criticism to observe that if our language is not here fully displayed, I have only failed in an attempt which no human powers have hitherto completed. If the lexicons of ancient tongues, now immutably fixed, and comprised in a few volumes, be yet, after the toil of successive ages, inadequate and delusive; if the aggregated knowledge and co-operating diligence of the Italian academicians did not secure them from the censure of Beni; if the embodied critics of France, when fifty years had been spent upon their work, were obliged to change its economy, and give their second edition another form, I may surely be contented without the praise of perfection, which, if I could obtain, in this gloom of solitude, what would it avail me? I have protracted my work till most of those whom I wished to please have sunk into the grave, and success and miscarriage are empty sounds: I therefore dismiss it with frigid tranquillity, having little to fear or hope from censure or from praise.

from *An Essay on Criticism*

Thus Critics, of less judgment than caprice,
Curious not knowing, not exact but nice,
Form short Ideas; and offend in arts
(As most in manners) by a love to parts.
   Some to *Conceit* alone their taste confine,
And glitt'ring thoughts struck out at ev'ry line;
Pleas'd with a work where nothing's just or fit;
One glaring Chaos and wild heap of wit.
Poets like painters, thus, unskill'd to trace
The naked nature and the living grace,
With gold or jewels cover ev'ry part,
And hide with ornaments their want of art.
True Wit is Nature to advantage dress'd,
What oft was thought, but ne'er so well express'd;
Something, whose truth convinc'd at sight we find,
That gives us back the image of our mind.
As shades more sweetly recommend the light,
So modest plainness sets off sprightly wit.
For works may have more wit than does 'em good,
As bodies perish thro' excess of blood.
   Others for *Language* all their care express,
And value books, as women men, for Dress:
Their praise is still, — the Style is excellent:
The Sense, they humbly take upon content.
Words are like leaves; and where they most abound,
Much fruit of sense beneath is rarely found.
False Eloquence, like the Prismatic glass,
Its gaudy colours spreads on ev'ry place;
The face of nature we no more survey,
All glares alike, without distinction gay:
But true Expression, like th' unchanging Sun,
Clears, and improves whate'er it shines upon,
It gilds all objects, but it alters none.
Expression is the dress of thought, and still
Appears more decent, as more suitable;
A vile conceit in pompous words express'd,
Is like a clown in regal purple dress'd:
For diff'rent styles with diff'rent subjects sort,
As several garbs with country, town, and court.
Some by old words to fame have made pretence,

Ancients in phrase, meer moderns in their sense:
Such labour'd nothings, in so strange a style,
Amaze th' unlearn'd, and make the learned smile;
Unlucky, as Fungoso in the Play,[1]
These sparks with aukward vanity display
What the fine gentleman wore yesterday;
And but so mimic ancient wits at best,
As apes our grandsires, in their doublets drest.
In words, as fashions, the same rule will hold;
Alike fantastic, if too new, or old;
Be not the first by whom the new are try'd,
Nor yet the last to lay the old aside.
    But most by Numbers judge a Poet's song,
And smooth or rough, with them, is right or wrong;
In the bright Muse tho' thousand charms conspire,
Her Voice is all these tuneful fools admire;
Who haunt Parnassus[2] but to please their ear,
Not mend their minds; as some to Church repair,
Not for the doctrine, but the music there.
These equal syllables alone require,
Tho' oft the ear the open vowels tire;
While expletives their feeble aid do join;
And ten low words oft creep in one dull line;
While they ring round the same unvary'd chimes,
With sure returns of still expected rhymes.
Where-e'er you find 'the cooling western breeze,'
In the next line, it 'whispers thro' the trees;'
If crystal streams 'with pleasing murmurs creep,'
The reader's threaten'd (not in vain) with 'sleep.'
Then, at the last and only couplet fraught
With some unmeaning thing they call a thought,
A needless Alexandrine[3] ends the song,
That, like a wounded snake, drags its slow length along.
Leave such to tune their own dull rhymes, and know
What's roundly smooth, or languishingly slow;
And praise the easy vigour of a line,
Where Denham's strength, and Waller's[4] sweetness join.
True ease in writing comes from art, not chance,
As those move easiest who have learn'd to dance.
'Tis not enough no harshness gives offence,
The sound must seem an Echo to the sense:
Soft is the strain when Zephyr gently blows,
And the smooth stream in smoother numbers flows;

But when loud surges lash the sounding shoar,
The hoarse, rough verse should like the torrent roar.
When Ajax strives, some rock's vast weight to throw,
The line too labours, and the words move slow;
Not so, when swift Camilla scours the plain,
Flies o'er th' unbending corn, and skims along the main.
Hear how Timotheus' vary'd lays surprize,
And bid alternate passions fall and rise!⁵
While, at each change, the son of Libyan Jove
Now burns with glory, and then melts with love;
Now his fierce eyes with sparkling fury glow,
Now sighs steal out, and tears begin to flow:
Persians and Greeks like turns of nature found,
And the World's victor stood subdu'd by Sound!
The pow'r of Music all our hearts allow,
And what Timotheus was, is DRYDEN now.

Alexander Pope (1688–1744)

SAMUEL JOHNSON

Johnson bases his criticism of the 'metaphysical' poets upon the Augustan concept of the universality of truth. His own balanced, antithetical, polysyllabic prose shows how he feels truth to be best expressed, but he is too intelligent a reader to judge entirely according to theoretical criteria.

from *Life of Cowley*

W it, like all other things subject by their nature to the choice of man, has its changes and fashions, and at different times takes different forms. About the beginning of the seventeenth century appeared a race of writers that may be termed the metaphysical poets, of whom in a criticism on the works of Cowley it is not improper to give some account.

The metaphysical poets were men of learning, and to show their learning was their whole endeavour; but, unluckily resolving to show it in rhyme, instead of writing poetry they only wrote verses, and very often such verses as stood the trial of the finger better than of the ear; for the modulation was so imperfect that they were only found to be verses by counting the syllables.

If the father of criticism[1] has rightly denominated poetry τέχνη μιμητική,[2] *an imitative art*, these writers will without great wrong lose their right to the name of poets, for they cannot be said to have imitated any thing: they neither copied nature nor life; neither painted the forms of matter nor represented the operations of intellect.

Those however who deny them to be poets allow them to be wits. Dryden confesses of himself and his contemporaries that they fall below Donne in wit, but maintains that they surpass him in poetry.

If wit be well described by Pope as being 'that which has been often thought, but was never before so well expressed',[3] they certainly never attained nor ever sought it, for they endeavoured to be singular in their thoughts, and were careless of their diction. But Pope's account of wit is undoubtedly erroneous; he depresses it below its natural dignity, and reduces it from strength of thought to happiness of language.

If by a more noble and more adequate conception that be considered as wit which is at once natural and new, that which though not obvious is, upon its first production, acknowledged to be just; if it be that which he that never found it wonders how he missed; to wit of this kind the metaphysical poets have seldom risen. Their thoughts are often new, but seldom natural; they are not obvious, but neither are they just; and the reader, far from wondering that he missed them, wonders more frequently by what perverseness of industry they were ever found.

But wit, abstracted from its effects upon the hearer, may be more rigorously and philosophically considered as a kind of *discordia concors,*[4] a combination of dissimilar images, or discovery of occult resemblances in things apparently unlike. Of wit, thus defined, they have more than enough. The most heterogenous ideas are yoked by violence together; nature and art are ransacked for illustrations, comparisons, and allusions; their learning instructs, and their subtlety surprises; but the reader commonly thinks his improvement dearly bought, and, though he sometimes admires, is seldom pleased.

From this account of their compositions it will be readily inferred that they were not successful in representing or moving the affections. As they were wholly employed on something unexpected and surprising they had no regard to that uniformity of sentiment which enables us to conceive and to excite the pains and the pleasure of other minds: they never enquired what on any occasion they should have said or done, but wrote rather as beholders than partakers of human nature; as beings looking upon good and evil, impassive and at leisure; as Epicurean[5] deities making remarks on the actions of men and the vicissitudes of life, without interest and without emotion. Their courtship was void of fondness and their lamentation of sorrow. Their wish was only to say what they hoped had been never said before.

Nor was the sublime more within their reach than the pathetic; for they never attempted that comprehension and expanse of thought which at once fills the whole mind, and of which the first effect is sudden astonishment, and the second rational admiration. Sublimity is produced by aggregation, and littleness by dispersion. Great thoughts are always general, and consist in positions not limited by exceptions, and in descriptions not descending to minuteness. It is with great propriety that subtlety, which in its original import means exility of particles,[6] is taken in its metaphorical meaning for nicety of distinction. Those writers who lay on the watch for novelty could have little hope of greatness; for great things cannot have escaped former observation. Their attempts were always analytic: they broke every image into fragments, and could no more represent by their slender conceits and laboured particularities the prospects of nature or the scenes of life than he who dissects a sunbeam with a prism can exhibit the wide effulgence of a summer noon.

What they wanted however of the sublime they endeavoured to supply by hyperbole; their amplification had no limits: they left not only reason but fancy behind them, and produced combinations of confused magnificence that not only could not be credited, but could not be imagined.

Yet great labour directed by great abilities is never wholly lost: if they frequently threw away their wit upon false conceits, they likewise sometimes struck out unexpected truth: if their conceits were far-fetched, they were often worth the carriage. To write on their plan it was at least necessary to read and think. No man could be born a metaphysical poet, nor assume the dignity of a writer by descriptions copied from descriptions, by imitations borrowed from imitations, by traditional imagery and hereditary similes, by readiness of rhyme and volubility of syllables.

In perusing the works of this race of authors the mind is exercised either by recollection or inquiry; either something already learned is to be retrieved, or something new is to be examined. If their greatness seldom elevates, their acuteness often surprises; if the imagination is not always gratified, at least the powers of reflection and comparison are employed; and in the mass of materials, which ingenious absurdity has thrown together, genuine wit and useful knowledge may be sometimes found, buried perhaps in grossness of expression, but useful to those who know their value, and such as, when they are expanded to perspicuity and polished to elegance, may give lustre to works which have more propriety though less copiousness of sentiment.

from *The Task*

There is a pleasure in poetic pains
Which only poets know. The shifts and turns,
The expedients and inventions multiform
To which the mind resorts, in chase of terms
Though apt, yet coy, and difficult to win, —
To arrest the fleeting images that fill
The mirror of the mind, and hold them fast,
And force them sit, till he has pencil'd off
A faithful likeness of the forms he views;
Then to dispose his copies with such art
That each may find its most propitious light,
And shine by situation, hardly less
Than by the labour and the skill it cost,
Are occupations of the poet's mind
So pleasing, and that steal away the thought
With such address, from themes of sad import,
That lost in his own musings, happy man!
He feels the anxieties of life, denied
Their wonted entertainment, all retire.
Such joys has he that sings. But ah! not such,
Or seldom such, the hearers of his song.
Fastidious, or else listless, or perhaps
Aware of nothing arduous in a task
They never undertook, they little note
His dangers or escapes, and haply find
There least amusement where he found the most.
But is amusement all? studious of song,
And yet ambitious not to sing in vain,
I would not trifle merely, though the world
Be loudest in their praise who do no more.
Yet what can satire, whether grave or gay?
It may correct a foible, may chastise
The freaks of fashion, regulate the dress,
Retrench a sword-blade, or displace a patch;
But where are its sublimer trophies found?
What vice has it subdued? whose heart reclaim'd
By rigour, or whom laugh'd into reform?
Alas! Leviathan[1] is not so tamed.

Laugh'd at, he laughs again; and stricken hard,
Turns to the stroke his adamantine scales,
That fear no discipline of human hands.

William Cowper (1731–1800)

EDWARD GIBBON

Edward Gibbon (1737–94) entered Parliament in 1774, but is chiefly remembered for his *History of the Decline and Fall of the Roman Empire*, which took sixteen years to write. His *Memoirs*, edited by the Lord Sheffield mentioned in this extract, were published in 1796. He reveals a good deal about his research methods, his consciousness of style, and much more besides.

from *Memoirs of my Life and Writings*

In the fifteen years between my *Essay on the Study of Literature* and the first volume of the *Decline and Fall* (1761–76), this criticism on Warburton, and some articles in the Journal, were my sole publications. It is more specially incumbent on me to mark the employment, or to confess the waste of time, from my travels to my father's death, an interval in which I was not diverted by any professional duties from the labours and pleasures of a studious life. (1) As soon as I was released from the fruitless task of the Swiss revolutions (1768), I began gradually to advance from the wish to the hope, from the hope to the design, from the design to the execution, of my historical work, of whose limits and extent I had yet a very inadequate notion. The Classics, as low as Tacitus, the younger Pliny, and Juvenal,[1] were my old and familiar companions. I insensibly plunged into the ocean of the Augustan history; and in the descending series I investigated, with my pen almost always in my hand, the original records, both Greek and Latin, from Dion Cassius to Ammianus Marcellinus,[2] from the reign of Trajan to the last age of the Western Caesars. The subsidiary rays of medals, and inscriptions of geography and chronology, were thrown on their proper objects; and I applied the collections of Tillemont,[3] whose inimitable accuracy almost assumes the character of genius, to fix and arrange within my reach the loose and scattered atoms of historical information. Through the darkness of the middle ages I explored my way in the Annals and Antiquities of Italy of the learned Muratori; and diligently compared them with the parallel or transverse lines of Sigonius and Maffei, Baronius and Pagi,[4] till I almost grasped the ruins of Rome in the fourteenth century, without suspecting that this final chapter must be attained by the labour of six quartos and twenty years. Among the books which I purchased, the Theodosian Code, with the commentary of James Godefroy,[5] must be gratefully remembered; I used it (and much I used it) as a work of history, rather than of jurisprudence: but in every light it may be considered as a full and capacious repository of the political state of the empire in the fourth and fifth centuries. As I believed, and as I still believe, that the propagation of the Gospel, and the triumph of the Church, are inseparably connected with the decline of the Roman monarchy, I weighed the causes and effects of the revolution, and contrasted the narratives and apologies of the Christians themselves, with the glances of candour or enmity which the Pagans have cast on the rising sects. The Jewish and heathen testimonies, as they are collected and illustrated by Dr Lardner,[6] directed, without superseding, my search of the originals; and in an ample dissertation on the miraculous darkness of the passion, I privately drew my conclusions from the silence of an unbelieving age. I have assembled the preparatory studies, directly or indirectly relative to my history; but, in strict equity, they must be spread beyond this period of my life, over the two summers (1771 and 1772) that elapsed between my father's death and my settlement in London. (2) In a free conversation with books and men, it would be endless to enumerate the names and characters of all who are introduced to our acquaintance; but in this general acquaintance we may select the degrees of friendship and esteem. According to the wise maxim, *Multum legere potius quam multa*,[7] I reviewed, again and again, the immortal works of the French and English, the Latin and Italian classics. My Greek studies (though less assiduous than I designed) maintained and extended my knowledge of that incomparable idiom. Homer and Xenophon were still my favourite authors; and I had almost prepared for the press an *Essay on the Cyropaedia*, which, in my own judgement, is not unhappily laboured. After a certain age, the new publications of merit are the sole food of the many; and the most austere student will often be tempted to break the line, for the sake of indulging his own curiosity, and of providing the topics of fashionable currency. A more respectable motive may be assigned for the triple perusal of Blackstone's *Commentaries*,[8] and a copious and critical abstract of that English work was my first serious production in my native language. (3) My literary leisure was much less complete and independent than it might appear to the eye of a stranger. In the hurry of London I was destitute of books; in the solitude of Hampshire I was not master of my time. My quiet was gradually disturbed by our domestic anxiety, and I should be ashamed of my unfeeling philosophy had I found much time to waste for study in the last fatal summer (1770) of my father's decay and dissolution.

The disembodying of the militia at the close of the war (1763) had restored the Major (a new Cincinnatus) to a life of agriculture. His labours were useful, his pleasures innocent, his wishes moderate; and my father *seemed* to enjoy the state of happiness which is celebrated by poets and philosophers, as the most agreeable to nature, and the least accessible to fortune.

But the last indispensable condition, the freedom from debt, was wanting to my father's felicity; and the vanities of his youth were severely punished by the solicitude and sorrow of his declining age. The first mortgage, on my return from Lausanne (1758), had afforded him a partial and transient relief.

The annual demand of interest and allowance was a heavy deduction from his income; the militia was a source of expense, the farm in his hands was not a profitable adventure, he was loaded with the costs and damages of an obsolete lawsuit; and each year multiplied the number, and exhausted the patience, of his creditors. Under these painful circumstances my own behaviour was not only guiltless but meritorious. Without stipulating any personal advantages, I consented, at a mature and well-informed age, to an additional mortgage, to the sale of Putney, and to every sacrifice that could alleviate his distress. But he was no longer capable of a rational effort, and his reluctant delays postponed not the evils themselves, but the remedies of those evils. The pangs of shame, tenderness, and self-reproach, incessantly preyed on his vitals; his constitution was broken; he lost his strength and his sight: the rapid progress of a dropsy admonished him of his end, and he sunk into the grave on the 10th of November, 1770, in the sixty-fourth year of his age. A family tradition insinuates that Mr William Law has drawn his pupil in the light and inconstant character of *Flatus*,[9] who is ever confident, and ever disappointed in the chase of happiness. But these constitutional failings were amply compensated by the virtues of the head and heart, by the warmest sentiments of honour and humanity. His graceful person, polite address, gentle manners, and unaffected cheerfulness, recommended him to the favour of every company; and in the change of times and opinions, his liberal spirit had long since delivered him from the zeal and prejudice of a Tory education. The tears of a son are seldom lasting: I submitted to the order of nature; and my grief was soothed by the conscious satisfaction that I had discharged all the duties of filial piety. Few, perhaps, are the children who, after the expiration of some months or years, would sincerely rejoice in the resurrection of their parents; and it is a melancholy truth that my father's death, not unhappy for himself, was the only event that could save me from an hopeless life of obscurity and indigence.

As soon as I had paid the last solemn duties to my father, and obtained, from time and reason, a tolerable composure of mind, I began to form the plan of an independent life most adapted to my circumstances and inclination. Yet so intricate was the net, my efforts were so awkward and feeble, that near two years (November, 1770–October 1772) were suffered to elapse before I could disentangle myself from the management of the farm, and transfer my residence from Buriton to a house in London. During this interval I continued to divide my year between town and the country; but my new freedom was brightened by hope; my stay in London was prolonged into the summer; and the uniformity of the summer was occasionally broken by visits and excursions at a distance from home. The gratification of my desires (they were not immoderate) has been seldom disappointed by the want of money or credit; my pride was never insulted by the visit of an importunate tradesman; and any transient anxiety for the past or future was soon dispelled by the studious or social occupation of the present hour. My conscience does not accuse me of any act of extravagance or injustice; the remnant of my estate affords an ample and honourable provision for my declining age. I shall not expatiate more minutely on my economical affairs, which cannot be instructive or amusing to the reader. It is a rule of prudence, as well as of politeness, to reserve such confidence for the ear of a private friend, without exposing our situation to the envy or pity of strangers; for envy is productive of hatred, and pity borders too nearly on contempt. Yet I may believe, and even assert, that in circumstances more indigent or more wealthy, I should never have accomplished the task, or acquired the fame, of an historian; that my spirit would have been broken by poverty and contempt, and that my industry might have been relaxed in the labour and luxury of a superfluous fortune. Few works of merit and importance have been executed either in a garret or a palace. A gentleman possessed of leisure and independence, of books and talents, may be encouraged to write by the distant prospect of honour and reward: but wretched is the author, and wretched will be the work, where daily diligence is stimulated by daily hunger.

I had now attained the first of earthly blessings, independence: I was the absolute master of my hours and actions: nor was I deceived in the hope that the establishment of my library in town would allow me to divide the day between study and society. Each year the circle of my acquaintance, the number of my dead and living companions, was enlarged. To a lover of books, the shops and sales of London present irresistible temptations; and the manufacture of my History required a various and growing stock of materials. The militia, my travels, the House of Commons, the fame of an author contributed to multiply my connexions: I was chosen a member of the fashionable clubs; and, before I left England in 1783, there were few persons of any eminence in the literary or political world to whom I was a stranger. It would most assuredly be in my power to amuse the reader with a gallery of portraits and a collection of anecdotes. But I have always condemned the practice of transforming a private memorial into a vehicle of satire or praise. By my own choice I passed in town the greatest part of the year; but whenever I was desirous of breathing the air of the country, I possessed an hospitable retreat at Sheffield Place in Sussex, in the family of my valuable friend Mr Holroyd, whose character, under the name of Lord Sheffield, has since been more conspicuous to the public.

No sooner was I settled in my house and library, than I undertook the composition of the first volume of my *History*. At the outset all was dark and doubtful; even the title of the work, the true era of the *Decline and Fall of the Empire*, the limits of the introduction, the division of the chapters, and the order of the narrative; and I was often tempted to cast away the labour of seven years. The style of an author should be the image of his mind, but the choice and command of language is the fruit of exercise. Many experiments were made before I could hit the middle tone between a dull chronicle and a rhetorical declamation: three times did I compose the first chapter, and twice the second and third, before I was tolerably satisfied with their effect. In the remainder of the way I advanced with a more equal and easy pace; but the fifteenth and sixteenth chapters have been reduced by three successive revisals from a large volume to their present size; and they might still be compressed, without any loss of facts or sentiments. An opposite fault may be imputed to the concise and superficial narrative of the first reigns from Commodus to Alexander; a fault of which I have never heard, except from Mr Hume in his last journey to London.[10] Such an oracle might have been consulted and obeyed with rational devotion; but I was soon disgusted with the modest practice of reading the manuscript to my friends. Of such friends some will praise from politeness, and some will criticize from vanity. The author himself is the best judge of his own performance; none has so deeply meditated on the subject; none is so sincerely interested in the event.

from *The Marriage of Heaven and Hell*

The Voice of the Devil

All Bibles or sacred codes have been the causes of the following Errors: —

1. That Man has two real existing principles, viz. a Body and a Soul.

2. That Energy, call'd Evil, is alone from the Body; and that Reason, call'd Good, is alone from the Soul.

3. That God will torment Man in Eternity for following his Energies.

But the following Contraries to these are True: —

1. Man has no Body distinct from his Soul; for that call'd Body is a portion of Soul discern'd by the five Senses, the chief inlets of Soul in this age.

2. Energy is the only life, and is from the Body; and Reason is the bound or outward circumference of Energy.

3. Energy is Eternal Delight.

Those who restrain Desire, do so because theirs is weak enough to be restrained; and the restrainer or Reason usurps its place and governs the unwilling.

And being restrained, it by degrees becomes passive, till it is only the shadow of Desire.

The history of this is written in *Paradise Lost*, and the Governor or Reason is call'd Messiah.

And the original Archangel, or possessor of the command of the Heavenly Host, is call'd the Devil or Satan, and his children are call'd Sin and Death.

But in the book of Job, Milton's Messiah is called Satan.

For this history has been adopted by both parties.

It indeed appear'd to Reason as if Desire was cast out; but the Devil's account is, that the Messiah fell, and formed a Heaven of what he stole from the Abyss.

This is shown in the Gospel, where he prays to the Father to send the Comforter, or Desire, that Reason may have Ideas to build on; the Jehovah of the Bible being no other than he who dwells in flaming fire.

Know that after Christ's death, he became Jehovah.

But in Milton, the Father is Destiny, the Son a Ratio of the five senses, and the Holy-ghost Vacuum!

Note. The reason Milton wrote in fetters when he wrote of Angels and God, and at liberty when of Devils and Hell, is because he was a true Poet, and of the Devil's party without knowing it.

from *Milton*

And did those feet in ancient time
   Walk upon England's mountains green?
And was the holy Lamb of God
   On England's pleasant pastures seen?

And did the Countenance Divine
   Shine forth upon our clouded hills?
And was Jerusalem builded here
   Among these dark Satanic Mills?

Bring me my bow of burning gold!
   Bring me my arrows of desire!
Bring me my spear! O clouds, unfold!
   Bring me my chariot of fire!

I will not cease from mental fight,
   Nor shall my sword sleep in my hand,
Till we have built Jerusalem
   In England's green and pleasant land.

William Blake (1757–1827)

JAMES BOSWELL

James Boswell (1740–95) derives his reputation largely from his talent and industry in observing and reporting the words and actions of Dr Johnson. The discussion of linguistic and other matters in this extract exemplifies Boswell's apparently self-effacing manner, though since the selection of topic and diction is entirely his, the narrative must tell us at least as much about him as it does about Johnson.

from *The Life of Johnson*

On Monday, March 23, I found him busy, preparing a fourth edition of his folio Dictionary. Mr. Peyton, one of his original amanuenses, was writing for him. I put him in mind of a meaning of the word *side*, which he had omitted, viz. relationship; as father's side, mother's side. He inserted it. I asked him if *humiliating* was a good word. He said, he had seen it frequently used, but he did not know it to be legitimate English. He would not admit *civilization*, but only *civility*. With great deference to him, I thought *civilization*, from *to civilize* better in the sense opposed to *barbarity*, than *civility*; as it is better to have a distinct word for each sense, than one word with two senses, which *civility* is, in his way of using it.

He seemed also to be intent on some sort of chymical operation. I was entertained by observing how he contrived to send Mr. Peyton on an errand, without seeming to degrade him. 'Mr. Peyton, — Mr. Peyton, will you be so good as to take a walk to Temple-Bar? You will there see a chymist's shop; at which you will be pleased to buy for me an ounce of oil of vitriol; not spirit of vitriol, but oil of vitriol. It will cost three halfpence.' Peyton immediately went, and returned with it, and told him it cost but a penny.

I then reminded him of the schoolmaster's cause,[1] and proposed to read to him the printed papers concerning it. 'No, Sir, (said he,) I can read quicker than I can hear.' So he read them to himself.

After he had read for some time, we were interrupted by the entrance of Mr. Kristrom, a Swede, who was tutor to some young gentlemen in the city. He told me, that there was a very good History of Sweden, by Daline. Having at that time an intention of writing the history of that country, I asked Dr. Johnson whether one might write a history of Sweden, without going thither. 'Yes, Sir, (said he,) one for common use.'

We talked of languages. Johnson observed, that Leibnitz had made some progress in a work, tracing all languages up to the Hebrew. 'Why, Sir, (said he,) you would not imagine that the French *jour*, day, is derived from the Latin *dies*, and yet nothing is more certain; and the intermediate steps are very clear. From *dies*, comes *diurnus*. *Diu* is, by inaccurate ears, or inaccurate pronunciation, easily confounded with *giu*; then the Italians form a substantive of the ablative of an adjective, and thence *giurno*, or, as they make it, *giorno*; which is readily contracted into *giour*, or *jour*.' He observed, that the Bohemian language was true Sclavonick. The Swede said, it had some similarity with the German. JOHNSON. 'Why, Sir, to be sure, such parts of Sclavonia as confine with Germany, will borrow German words; and such parts as confine with Tartary will borrow Tartar words.'

He said, he never had it properly ascertained that the Scotch Highlanders and the Irish understood each other. I told him that my cousin Colonel Graham, of the Royal Highlanders, whom I met at Drogheda, told me they did. JOHNSON. 'Sir, if the Highlanders understood Irish, why translate the New Testament into Erse, as was done lately at Edinburgh, when there is an Irish translation?' BOSWELL. 'Although the Erse and Irish are both dialects of the same language, there may be a good deal of diversity between them, as between the different dialects in Italy.' — The Swede went away, and Mr. Johnson continued his reading of the papers. I said, 'I am afraid, Sir, it is troublesome to you.' 'Why, Sir, (said he,) I do not take much delight in it; but I'll go through with it.'

We went to the Mitre, and dined in the room where he and I first supped together. He gave me great hopes of my cause. 'Sir, (said he,) the government of a schoolmaster is somewhat in the nature of military government; that is to say, it must be arbitrary, it must be exercised by the will of one man, according to particular circumstances. You must shew some learning upon this occasion. You must shew, that a schoolmaster has a prescriptive right to beat; and that an action of assault and battery cannot be admitted against him, unless there is some great excess, some barbarity. This man has maimed none of his boys. They are all left with the full exercise of their corporeal faculties. In our schools in England, many boys have been maimed; yet I never heard of an action against a schoolmaster on that account. Puffendorf, I think, maintains the right of a schoolmaster to beat his scholars.'

On Saturday, March 27 [*sic.*], I introduced him Sir Alexander Macdonald, with whom he had expressed a wish to be acquainted. He received him very courteously. . . .

SIR A. 'I have been correcting several Scotch accents in my friend Boswell. I doubt, Sir, if any Scotchman ever attains to a perfect English pronunciation.' JOHNSON. 'Why, Sir, few of them do, because they do not persevere after acquiring a certain degree of it. But, Sir, there can be no doubt that they may attain to a perfect English pronunciation, if they will. We find how near they come to it; and certainly, a man who conquers nineteen parts of the Scottish accent, may conquer the twentieth. But, Sir, when a man has got the better of nine tenths, he grows weary, he relaxes his diligence, he finds he has corrected his accent so far as not to be disagreeable, and he no longer desires his friends to tell him when he is wrong; nor does he choose to be told. Sir, when people watch me narrowly, and I do not watch myself, they will find me out to be of a particular county. In the same manner, Dunning may be found out to be a Devonshire man. So most Scotchmen may be found out. But, Sir, little aberrations are of no disadvantage. I never catched Mallet in a Scotch accent; and yet Mallet, I suppose, was past five-and-twenty before he came to London.'

Upon another occasion I talked to him on this subject, having myself taken some pains to improve my pronunciation, by the aid of the late Mr. Love, of Drury-lane theatre, when he was a player at Edinburgh, and also of old Mr Sheridan. Johnson said to me, 'Sir, your pronunciation is not offensive.' With this concession I was pretty well satisfied; and let me give my countrymen of North-Britain an advice not to aim at absolute perfection in this respect; not to speak *High English*, as we are apt to call what is far removed from the *Scotch*, but which is by no means *good English*, and makes, 'the fools who use it,' truly ridiculous. Good English is plain, easy, and smooth in the mouth of an unaffected English Gentleman. A studied and factitious pronunciation, which requires perpetual attention and imposes perpetual constraint, is exceedingly disgusting. A small intermixture of provincial peculiarities may, perhaps, have an agreeable effect, as the notes of different birds concur in the harmony of the grove, and please more than if they were all exactly alike. I could name some gentlemen of Ireland, to whom a slight proportion of the accent and recitative of that country is an advantage. The same observation will apply to the gentlemen of Scotland. I do not mean that we should speak as broad as a certain prosperous member of Parliament from that country; though it has been well observed, that 'it has been of no small use to him; as it rouses the attention of the House by its uncommonness; and is equal to tropes and figures in a good English speaker.' I would give as an instance of what I mean to recommend to my countrymen, the pronunciation of the late Sir Gilbert Elliot; and may I presume to add that of the present Earl of Marchmont, who told me, with great good humour, that the master of a shop in London, where he was not known, said to him, 'I suppose, Sir, you are an American.' 'Why so, Sir?' (said his Lordship.) 'Because, Sir, (replied the shopkeeper,) you speak neither English nor Scotch, but something different from both, which I conclude is the language of America.'

BOSWELL. 'It may be of use, Sir, to have a Dictionary to ascertain the pronunciation.' JOHNSON. 'Why, Sir, my Dictionary shows you the accents of words, if you can but remember them.' BOSWELL. 'But, Sir, we want marks to ascertain the pronunciation of the vowels. Sheridan, I believe, has finished such a work.' JOHNSON. 'Why, Sir, consider how much easier it is to learn a language by the ear, than by any marks. Sheridan's Dictionary may do very well; but you cannot always carry it about with you; and, when you want the word, you have not the Dictionary. It is like a man who has a sword that will not draw. It is an admirable sword, to be sure: but while your enemy is cutting your throat, you are unable to use it. Besides, Sir, what entitles Sheridan to fix the pronunciation of English? He has, in the first place, the disadvantage of being an Irishman: and if he says he will fix it after the example of the best company, why, they differ among themselves. I remember an instance: when I published the Plan for my Dictionary, Lord Chesterfield told me that the word *great* should be pronounced so as to rhyme to *state*; and Sir William Yonge sent me word that it should be pronounced so as to rhyme to *seat*, and that none but an Irishman would pronounce it *grait*. Now here were two men of the highest rank, the one, the best speaker in the House of Lords, the other, the best speaker in the House of Commons, differing entirely.'

Nuns Fret Not at Their Convent's Narrow Room

Nuns fret not at their convent's narrow room;
And hermits are contented with their cells;
And students with their pensive citadels;
Maids at the wheel, the weaver at his loom,
Sit blithe and happy; bees that soar for bloom,
High as the highest Peak of Furness-fells,
Will murmur by the hour in foxglove bells:
In truth the prison, unto which we doom
Ourselves, no prison is: and hence for me,
In sundry moods, 'twas pastime to be bound
Within the Sonnet's scanty plot of ground;
Pleased if some Souls (for such there needs must be)
Who have felt the weight of too much liberty,
Should find brief solace there, as I have found.

William Wordsworth (1770–1850)

JANE AUSTEN

Jane Austen (1775–1817) wove her quiet English middle-class country background into six novels of genius. Her first, *Northanger Abbey*, remained unpublished until after her death. The extract below represents one of the few occasions on which she appears to address the reader directly in order to reveal her views about the craft and achievement of the creative writer.

from *Northanger Abbey*

T he progress of the friendship between Catherine and Isabella was quick as its beginning had been warm, and they passed so rapidly through every gradation of increasing tenderness, that there was shortly no fresh proof of it to be given to their friends or themselves. They called each other by their Christian name, were always arm in arm when they walked, pinned up each other's train for the dance, and were not to be divided in the set; and if a rainy morning deprived them of other enjoyments, they were still resolute in meeting in defiance of wet and dirt, and shut themselves up, to read novels together. Yes, novels; — for I will not adopt that ungenerous and impolitic custom so common with novel writers, of degrading by their contemptuous censure the very performances, to the number of which they are themselves adding — joining with their greatest enemies in bestowing the harshest epithets on such works, and scarcely ever permitting them to be read by their own heroine, who, if she accidentally take up a novel, is sure to turn over its insipid pages with disgust. Alas! if the heroine of one novel be not patronized by the heroine of another, from whom can she expect protection and regard? I cannot approve of it. Let us leave it to the Reviewers to abuse such effusions of fancy at their leisure, and over every new novel to talk in threadbare strains of the trash with which the press now groans. Let us not desert one another; we are an injured body. Although our productions have afforded more extensive and unaffected pleasure than those of any other literary corporation in the world, no species of composition has been so much decried. From pride, ignorance, or fashion, our foes are almost as many as our readers. And while the abilities of the nine-hundredth abridger of the History of England, or of the man who collects and publishes in a volume some dozen lines of Milton, Pope, and Prior, with a paper from the Spectator, and a chapter from Sterne, are eulogized by a thousand pens, — there seems almost a general wish of decrying the capacity and undervaluing the labour of the novelist, and of slighting the performances which have only genius, wit, and taste to recommend them. 'I am no novel reader — I seldom look into novels — Do not imagine that *I* often read novels — It is really very well for a novel.' — Such is the common cant. — 'And what are you reading, Miss —— ?' 'Oh! it is only a novel!' replies the young lady; while she lays down her book with affected indifference, or momentary shame. — 'It is only Cecilia, or Camilla, or Belinda;' or, in short, only some work in which the greatest powers of the mind are displayed, in which the most thorough knowledge of human nature, the happiest delineation of its varieties, the liveliest effusions of wit and humour are conveyed to the world in the best chosen language. Now, had the same young lady been engaged with a volume of the Spectator, instead of such a work, how proudly she would have produced the book, and told its name; though the chances must be against her being occupied by any part of that voluminous publication, of which either the matter or manner would not disgust a young person of taste: the substance of its papers so often consisting in the statement of improbable circumstances, unnatural characters, and topics of conversation, which no longer concern any one living; and their language, too, frequently so coarse as to give no very favourable idea of the age that could endure it.

from *Don Juan*

My poem's epic, and is meant to be
   Divided in twelve books; each book containing,
With love, and war, a heavy gale at sea,
   A list of ships, and captains, and kings reigning,
New characters; the episodes are three:
   A panoramic view of hell's in training,
After the style of Virgil and of Homer,
So that my name of Epic's no misnomer.

All these things will be specified in time,
   With strict regard to Aristotle's rules,
The *Vade Mecum* of the true sublime,
   Which makes so many poets, and some fools:
Prose poets like blank-verse, I'm fond of rhyme,
   Good workmen never quarrel with their tools;
I've got new mythological machinery,
And very handsome supernatural scenery.

There's only one slight difference between
   Me and my epic brethren gone before,
And here the advantage is my own, I ween
   (Not that I have not several merits more,
But this will more peculiarly be seen);
   They so embellish, that 'tis quite a bore
Their labyrinth of fables to thread through,
Whereas this story's actually true.

If any person doubt it, I appeal
   To history, tradition, and to facts,
To newspapers, whose truth all know and feel,
   To plays in five, and operas in three acts;
All these confirm my statement a good deal,
   But that which more completely faith exacts
Is, that myself, and several now in Seville,
Saw Juan's last elopement with the devil.

If ever I should condescend to prose,
   I'll write poetical commandments, which
Shall supersede beyond all doubt all those
   That went before; in these I shall enrich

My text with many things that no one knows,
  And carry precept to the highest pitch:
I'll call the work 'Longinus[1] o'er a Bottle,
Or, Every Poet his *own* Aristotle.'

Thou shalt believe in Milton, Dryden, Pope;
  Thou shalt not set up Wordsworth, Coleridge, Southey;
Because the first is crazed beyond all hope,
  The second drunk, the third so quaint and mouthy:
With Crabbe it may be difficult to cope,
  And Campbell's Hippocrene is somewhat drouthy:
Thou shalt not steal from Samuel Rogers, nor
Commit — flirtation with the muse of Moore.

George Gordon, Lord Byron (1788–1824)

WILLIAM HAZLITT

William Hazlitt (1778–1830), friend of Coleridge, Wordsworth, and Charles Lamb, wrote essays on a wide variety of subjects, notably on the Elizabethan drama. His championship of the *plain style* reflects the writer's constant need to reassess the capacities of the available language, a need often resulting in a vigorous rejection of the style and values of the previous generation.

On Familiar Style

It is not easy to write a familiar style. Many people mistake a familiar for a vulgar style, and suppose that to write without affectation is to write at random. On the contrary, there is nothing that requires more precision, and, if I may so say, purity of expression, than the style I am speaking of. It utterly rejects not only all unmeaning pomp, but all low, cant phrases, and loose, unconnected, *slipshod* allusions. It is not to take the first word that offers, but the best word in common use; it is not to throw words together in any combinations we please, but to follow and avail ourselves of the true idiom of the language. To write a genuine familiar or truly English style, is to write as any one would speak in common conversation, who had a thorough command and choice of words, or who could discourse with ease, force, and perspicuity, setting aside all pedantic and oratorical flourishes. Or to give another illustration, to write naturally is the same thing in regard to common conversation, as to read naturally is in regard to common speech. It does not follow that it is an easy thing to give the true accent and inflection to the words you utter, because you do not attempt to rise above the level of ordinary life and colloquial speaking. You do not assume indeed the solemnity of the pulpit, or the tone of stage-declamation: neither are you at liberty to gabble on at a venture, without emphasis or discretion, or to resort to vulgar dialect or clownish pronunciation. You must steer a middle course. You are tied down to a given and appropriate articulation, which is determined by the habitual associations between sense and sound, and which you can only hit by entering into the author's meaning, as you must find the proper words and style to express yourself by fixing your thoughts on the subject you have to write about. Any one may mouth out a passage with a theatrical cadence, or get upon stilts to tell his thoughts: but to write or speak with propriety and simplicity is a more difficult task. Thus it is easy to affect a pompous style, to use a word twice as big as the thing you want to express: it is not so easy to pitch upon the very word that exactly fits it. Out of eight or ten words equally common, equally intelligible, with nearly equal pretensions, it is a matter of some nicety and discrimination to pick out the very one, the preferableness of which is scarcely perceptible, but decisive. The reason why I object to Dr. Johnson's style is, that there is no discrimination, no selection, no variety in it. He uses none but 'tall, opaque words,' taken from 'the first row of the rubric:' [1] — words with the greatest number of syllables, or Latin phrases with merely English terminations. If a fine style depended on this sort of arbitrary pretension, it would be fair to judge of an author's elegance by the measurement of his words, and the substitution of foreign circumlocutions (with no precise associations) for the mother-tongue.* How simple it is to be dignified without ease, to be pompous without meaning! Surely, it is but a mechanical rule for avoiding what is low to be always pedantic and affected. It is clear you cannot use a vulgar English word, if you never use a common English word at all. A fine tact is shewn in adhering to those which are perfectly common, and yet never falling into any expressions which are debased by disgusting circumstances, or which owe their signification and point to technical or professional allusions. A truly natural or familiar style can never be quaint or vulgar, for this reason, that it is of universal force and applicability, and that quaintness and vulgarity arise out of the immediate connection of certain words with coarse and disagreeable, or with confined ideas. The last form what we understand by *cant* or *slang* phrases. — To give an example of what is not very clear in the general statement. I should say that the phrase *To cut with a knife*, or *To cut a piece of wood*, is perfectly free from vulgarity, because it is perfectly common: but to *cut an acquaintance* is not quite unexceptionable, because it is not perfectly common or intelligible, and has hardly yet escaped out of the limits of slang phraseology. I should hardly therefore use the word in this sense without putting it in italics as a license of expression, to be received *cum grano salis*. All provincial or bye-phrases come under the same mark of reprobation — all such as the writer transfers to the page from his fire-side or a particular *coterie*, or that he invents for his own sole use and convenience. I conceive that words are like money, not the worse for being common, but that it is the stamp of custom alone that gives them circulation or value. I am fastidious in this respect, and would almost as soon coin the currency of the realm as counterfeit the King's English. I never invented or gave a new and unauthorised meaning to any word but one single one (the term *impersonal* applied to feelings) and that was in an abstruse metaphysical discussion to express a very difficult distinction. I have been (I know) loudly accused of revelling in vulgarisms and broken English. I cannot speak to that point: but so far I plead guilty to the determined use of acknowledged idioms and common elliptical expressions. I am not sure that the critics in question know the one from the other, that is, can distinguish any medium between

*I have heard of such a thing as an author, who makes it a rule never to admit a monosyllable into his vapid verse. Yet the charm and sweetness of Marlow's lines depended often on their being made up almost entirely of monosyllables.

formal pedantry and the most barbarous solecism. As an author, I endeavour to employ plain words and popular modes of construction, as were I a chapman and dealer, I should common weights and measures.

The proper force of words lies not in the words themselves, but in their application. A word may be a fine-sounding word, of an unusual length, and very imposing from its learning and novelty, and yet in the connection in which it is introduced, may be quite pointless and irrelevant. It is not pomp or pretension, but the adaptation of the expression to the idea that clenches a writer's meaning: — as it is not the size or glossiness of the materials, but their being fitted each to its place, that gives strength to the arch; or as the pegs and nails are as necessary to the support as the larger timbers, and more so than the mere shewy, unsubstantial ornaments. I hate any thing that occupies more space than it is worth. I hate to see a load of band-boxes go along the street, and I hate to see a parcel of big words without any thing in them. A person who does not deliberately dispose of all his thoughts alike in cumbrous draperies and flimsy disguises, may strike out twenty varieties of familiar everyday language, each coming somewhat nearer to the feeling he wants to convey, and at last not hit upon that particular and only one, which may be said to be identical with the exact impression in his mind. This would seem to shew that Mr. Cobbett is hardly right in saying that the first word that occurs is always the best.[2] It may be a very good one; and yet a better may present itself on reflection or from time to time. It should be suggested naturally, however, and spontaneously, from a fresh and lively conception of the subject. We seldom succeed by trying at improvement, or by merely substituting one word for another that we are not satisfied with, as we cannot recollect the name of a place or person by merely plaguing ourselves about it. We wander farther from the point by persisting in a wrong scent; but it starts up accidentally in the memory when we least expected it, by touching some link in the chain of previous association.

There are those who hoard up and make a cautious display of nothing but rich and rare phraseology; — ancient medals, obscure coins, and Spanish pieces of eight. They are very curious to inspect; but I myself would neither offer nor take them in the course of exchange. A sprinkling of archaisms is not amiss; but a tissue of obsolete expressions is more fit *for keep than wear.* I do not say I would not use any phrase that had been brought into fashion before the middle or the end of the last century; but I should be shy of using any that had not been employed by any approved author during the whole of that time. Words, like clothes, get old-fashioned, or mean and ridiculous, when they have been for some time laid aside. Mr. Lamb is the only imitator of old English style I can read with pleasure; and he is so thoroughly imbued with the spirit of his authors, that the idea of imitation is almost done away. There is an inward unction, a marrowy vein both in the thought and feeling, an intuition, deep and lively, of his subject, that carries off any quaintness or awkwardness arising from an antiquated style and dress. The matter is completely his own, though the manner is assumed. Perhaps his ideas are altogether so marked and individual, as to require their point and pungency to be neutralised by the affectation of a singular but traditional form of conveyance. Tricked out in the prevailing costume, they would probably seem more startling and out of the way. The old English authors, Burton, Fuller, Coryate, Sir Thomas Brown, are a kind of mediators between us and the more eccentric and whimsical modern, reconciling us to his peculiarities. I do not however know how far this is the case or not, till he condescends to write like one of us. I must confess that what I like best of his papers under the signature of Elia (still I do not presume, amid such excellence, to decide what is most excellent) is the account of *Mrs Battle's Opinions on Whist*, which is also the most free from obsolete allusions and turns of expression —

'A well of native English undefiled.' [3]

To those acquainted with his admired prototypes, these Essays of the ingenious and highly gifted author have the same sort of charm and relish, that Erasmus's Colloquies or a fine piece of modern Latin have to the classical scholar. Certainly, I do not know any borrowed pencil that has more power or felicity of execution than the one of which I have here been speaking.

It is as easy to write a gaudy style without ideas, as it is to spread a pallet of shewy colours, or to smear in a flaunting transparency. 'What do you read?' — 'Words, words, words.' — 'What is the matter?' — '*Nothing*,' it might be answered.[4] The florid style is the reverse of the familiar. The last is employed as an unvarnished medium to convey ideas; the first is resorted to as a spangled veil to conceal the want of them. When there is nothing to be set down but words, it costs little to have them fine. Look through the dictionary, and cull out a *florilegium*,[5] rival the *tulippomania*.[6] *Rouge* high enough, and never mind the natural complexion. The vulgar, who are not in the secret, will admire the look of preternatural health and vigour; and the fashionable, who regard only appearances, will be delighted with the imposition. Keep to your sounding generalities, your tinkling phrases, and all will be well. Swell out an unmeaning truism to a perfect tympany of style. A thought, a distinction is the rock on which all this brittle cargo of verbiage splits at once. Such writers have merely *verbal* imaginations, that retain nothing but words. Or their puny thoughts have dragon-wings, all green and gold. They soar far above the vulgar failing of the *Sermo humi obrepens*[7] — their most ordinary speech is never short of an hyperbole, splendid, imposing, vague, incomprehensible, magniloquent, a cento of sounding common-places. If some of us, whose 'ambition is more lowly,' [8] pry a little too narrowly into nooks and corners to pick up a number of 'unconsidered trifles',[9] they never once direct their eyes or lift their hands to seize on any but the most gorgeous, tarnished, thread-bare patch-work set of phrases, the left-off finery of poetic extravagance, transmitted down through successive generations of barren pretenders. If they criticise actors and actresses, a huddled phantasmagoria of feathers, spangles, floods of light, and oceans of sound float before their morbid sense, which they paint in the style of Ancient Pistol. Not a glimpse can you get of the merits or defects of the performers: they are hidden in a profusion of barbarous epithets and wilful rhodomontade. Our hypercritics are not thinking of these little fantoccini beings —

'That strut and fret their hour upon the stage'[10] —

but of tall phantoms of words, abstractions, *genera* and *species*, sweeping clauses, periods that unite the Poles, forced alliterations, astounding antitheses —

'And on their pens *Fustian* sits plumed.'[11]

If they describe kings and queens, it is an Eastern pageant. The Coronation at either House is nothing to it. We get at four repeated images — a curtain, a throne, a sceptre, and a foot-stool. These are with them the wardrobe of a lofty imagination; and they turn their servile strains to servile uses. Do we read a description of pictures? It is not a reflection of tones and hues which 'nature's own sweet and cunning hand laid on,'[12] but piles of precious stones, rubies, pearls, emeralds, Golconda's mines, and all the blazonry of art. Such persons are in fact besotted with words, and their brains are turned with the glittering, but empty and sterile phantoms of things. Personifications, capital letters, seas of sunbeams, visions of glory, shining inscriptions, the figures of a transparency, Britannia with her shield, or Hope leaning on an anchor, make up their stock in trade. They may be considered as *hieroglyphical* writers. Images stand out in their minds isolated and important merely in themselves, without any ground-work of feeling — there is no context in their imaginations. Words affect them in the same way, by the mere sound, that is, by their possible, not by their actual application to the subject in hand. They are fascinated by first appearances, and have no sense of consequences. Nothing more is meant by them than meets the ear: they understand or feel nothing more than meets their eye. The web and texture of the universe, and of the heart of man, is a mystery to them: they have no faculty that strikes a chord in unison with it. They cannot get beyond the daubings of fancy, the varnish of sentiment. Objects are not linked to feelings, words to things, but images revolve in splendid mockery, words represent themselves in their strange rhapsodies. The categories of such a mind are pride and ignorance — pride in outside show, to which they sacrifice every thing, and ignorance of the true worth and hidden structure both of words and things. With a sovereign contempt for what is familiar and natural, they are slaves of vulgar affectation — of a routine of high-flown phrases. Scorning to imitate realities, they are unable to invent any thing, to strike out one original idea. They are not copyists of nature, it is true: but they are the poorest of all plagiarists, the plagiarists of words. All is far-fetched, dear-bought, artificial, oriental in subject and allusion: all is mechanical, conventional, vapid, formal, pedantic in style and execution. They startle and confound the understanding of the reader, by the remoteness and obscurity of their illustrations: they soothe the ear by the monotony of the same everlasting round of circuitous metaphors. They are the *mock-school* in poetry and prose. They flounder about between fustian in expression, and bathos in sentiment. They tantalise the fancy, but never reach the head nor touch the heart. Their Temple of Fame is like a shadowy structure raised by Dulness to Vanity, or like Cowper's description of the Empress of Russia's palace of ice, as 'worthless as in shew 'twas glittering' —

'It smiled, and it was cold!' [13]

A Musical Instrument

What was he doing, the great god Pan,
    Down in the reeds by the river?
Spreading ruin and scattering ban,
Splashing and paddling with hoofs of a goat,
And breaking the golden lilies afloat
    With the dragon-fly on the river.

He tore out a reed, the great god Pan,
    From the deep cool bed of the river;
The limpid water turbidly ran,
And the broken lilies a-dying lay,
And the dragon-fly had fled away,
    Ere he brought it out of the river.

High on the shore sat the great god Pan,
    While turbidly flow'd the river ;
And hack'd and hew'd as a great god can
With his hard bleak steel at the patient reed,
Till there was not a sign of the leaf indeed
    To prove it fresh from the river.

He cut it short, did the great god Pan
    (How tall it stood in the river!)
Then drew the pith, like the heart of a man,
Steadily from the outside ring,
And notch'd the poor dry empty thing
    In holes, as he sat by the river.

'This is the way,' laugh'd the great god Pan
    (Laugh'd while he sat by the river),
'The only way, since gods began
To make sweet music, they could succeed.'
Then dropping his mouth to a hole in the reed,
    He blew in power by the river.

Sweet, sweet, sweet, O Pan!
    Piercing sweet by the river!
Blinding sweet, O great god Pan!
The sun on the hill forgot to die,
And the lilies revived, and the dragon-fly
    Came back to dream on the river.

Elizabeth Barrett Browning   85

Yet half a beast is the great god Pan,
 To laugh as he sits by the river,
Making a poet out of a man:
The true gods sigh for the cost and pain —
For the reed which grows nevermore again
 As a reed with the reeds of the river.

Elizabeth Barrett Browning (1806–61)

GEORGE BORROW

George Borrow (1803–81) claimed to know twelve languages by the time he was eighteen. He made his reputation with *The Bible in Spain*, based on his experiences as an agent of the British and Foreign Bible Society. His largely autobiographical novel *Lavengro* (which means 'word-master' in the Romany or Gypsy language) describes his early life and first steps as a writer.

from *Lavengro*

I crossed the river at a bridge considerably above that hight of [1] London; for, not being acquainted with the way, I missed the turning which should have brought me to the latter. Suddenly I found myself in a street of which I had some recollection, and mechanically stopped before the window of a shop at which various publications were exposed; it was that of the bookseller to whom I had last applied in the hope of selling my ballads or Ab Gwilym, and who had given me hopes that, in the event of my writing a decent novel or a tale, he would prove a purchaser. As I stood listlessly looking at the window, and the publications which it contained, I observed a paper affixed to the glass by wafers with something written upon it. I drew yet nearer for the purpose of inspecting it; the writing was in a fair round hand — 'A Novel or Tale is much wanted' was what was written. . . .

'I must do something,' said I, as I sat that night in my lonely apartment, with some bread and a pitcher of water before me.

Thereupon taking some of the bread, and eating it, I considered what I was to do. 'I have no idea what I am to do,' said I, as I stretched my hand towards the pitcher, 'unless — and here I took a considerable draught — I write a tale or a novel —— . That bookseller,' I continued, speaking to myself, 'is certainly much in need of a tale or novel, otherwise he would not advertise for one. Suppose I write one, I appear to have no other chance of extricating myself from my present difficulties; surely it was Fate that conducted me to his window.'

'I will do it,' said I, as I struck my hand against the table; 'I will do it.' Suddenly a heavy cloud of despondency came over me. Could I do it? Had I the imagination requisite to write a tale or a novel? 'Yes, yes,' said I, as I struck my hand again against the table, 'I can manage it; give me fair play, and I can accomplish anything.'

But should I have fair play? I must have something to maintain myself whilst I wrote my tale, and I had but eighteen pence in the world. Would that maintain me whilst I wrote my tale? Yes, I thought it would, provided I ate bread, which did not cost much, and drank water, which cost nothing;

it was poor diet, it was true, but better men than myself had written on bread and water; had not the big man told me so, or something to that effect, months before?

It was true there was my lodging to pay for; but up to the present time I owed nothing, and perhaps, by the time that the people of the house asked me for money, I should have written a tale or a novel, which would bring me in money; I had paper, pens, and ink, and, let me not forget them, I had candles in my closet, all paid for, to light me during my night work. Enough, I would go doggedly to work upon my tale or novel.

But what was the tale or novel to be about? Was it to be a tale of fashionable life, about Sir Harry Somebody, and the Countess Something? But I knew nothing about fashionable people, and cared less; therefore how should I attempt to describe fashionable life? What should the tale consist of? The life and adventures of some one. Good — but of whom? . . . I want a character for my hero, thought I, something higher than a mere robber; some one like — like Colonel B ——.[2] By the way, why should I not write the life and adventures of Colonel B —— of Londonderry, in Ireland?

A truly singular man was this same Colonel B —— of Londonderry, in Ireland; a personage of most strange and incredible feats and daring, who had been a partisan soldier, a bravo — who, assisted by certain discontented troopers, nearly succeeded in stealing the crown and regalia from the Tower of London; who attempted to hang the Duke of Ormonde at Tyburn; and whose strange, eventful career did not terminate even with his life, his dead body, on the circulation of an unfounded report that he did not come to his death by fair means, having been exhumed by the mob of his native place, where he had retired to die, and carried in the coffin through the streets.

Of his life I had inserted an account in the 'Newgate Lives and Trials';[3] it was bare and meagre, and written in the stiff, awkward style of the seventeenth century; it had, however, strongly captivated my imagination, and I now thought that out of it something better could be made; that, if I added to the adventures and purified the style, I might fashion out of it a very decent tale or novel. On a sudden, however, the proverb of mending old garments with new cloth occurred to me. 'I am afraid,' said I, 'any new adventures which I can invent will not fadge well with the old tale; one will but spoil the other.' I had better have nothing to do with Colonel B ——, thought I, but boldly and independently sit down and write the life of Joseph Sell.

This Joseph Sell, dear reader, was a fictitious personage who had just come into my head. I had never even heard of the name, but just at that moment it happened to come into my head; I would write an entirely fictitious narrative called the 'Life and Adventures of Joseph Sell, the Great Traveller.'

I had better begin at once, thought I; and, removing the bread and the jug, which latter was now empty, I seized pen and paper, and forthwith essayed to write the life of Joseph Sell, but soon discovered that it is much easier to resolve upon a thing than to achieve it, or even commence it; for the life of me I did not know how to begin, and, after trying in vain to write a line, I thought it would be as well to go to bed, and defer my projected undertaking till the morrow.

So I went to bed, but not to sleep. During the greater part of the night I lay awake, musing upon the work which I had determined to execute. For a long time my brain was dry and unproductive; I could form no plan which appeared feasible. At length I felt within my brain a kindly glow; it was the commencement of inspiration; in a few minutes I had formed my plan; I then began to imagine the scenes and the incidents. Scenes and incidents flitted before my mind's eye so plentifully that I knew not how to dispose of them; I was in a regular embarrassment. At length I got out of the difficulty in the easiest manner imaginable, namely, by consigning to the depths of oblivion all the feebler and less stimulant scenes and incidents, and retaining the better and more impressive ones. Before morning I had sketched the whole work on the tablets of my mind, and then resigned myself to sleep in the pleasing conviction that the most difficult part of my undertaking was achieved. . . .

Rather late in the morning I awoke; for a few minutes I lay still, perfectly still; my imagination was considerable sobered; the scenes and situations which had pleased me so much overnight appeared to me in a far less captivating guise that morning. I felt languid and almost hopeless — the thought, however, of my situation soon roused me — I must make an effort to improve the posture of my affairs; there was no time to be lost; so I sprang out of bed, breakfasted on bread and water, and then sat down doggedly to write the life of Joseph Sell.

It was a great thing to have formed my plan, and to have arranged the scenes in my head, as I had done the preceding night. The chief thing requisite at present was the mere mechanical act of committing them to paper. This I did not find so easy as I could wish — I wanted mechanical skill; but I persevered, and before evening I had written ten pages. I partook of some bread and water; and, before I went to bed that night, I had completed fifteen pages of my life of Joseph Sell.

The next day I resumed my task — I found my power of writing considerably increased; my pen hurried rapidly over the paper — my brain was in a wonderfully teeming state; many scenes and visions which I had not thought of before were evolved, and, as fast as evolved, written down; they seemed to be more pat to my purpose, and more natural to my history, than many others which I had imagined before, and which I made now give place to these newer creations: by about midnight I had added thirty fresh pages to my 'Life and Adventures of Joseph Sell.'

The third day arose — it was dark and dreary out of doors, and I passed it drearily enough within; my brain appeared to have lost much of its former glow, and my pen much of its power; I, however, toiled on, but at midnight had only added seven pages to my history of Joseph Sell.

On the fourth day the sun shone brightly. I arose, and, having breakfasted as usual, I fell to work. My brain was this day wonderfully prolific, and my pen never before or since glided so rapidly over the paper; towards night I began to feel strangely about the back part of my head, and my whole system was extraordinarily affected. I likewise occasionally saw double — a tempter now seemed to be at work within me.

'You had better leave off now for a short space,' said the tempter, 'and go out and drink a pint of beer; you still have one shilling left — if you go on at this rate, you will go mad — go out and spend sixpence, you can afford it, more than half your work is done.' I was about to obey the suggestion of the tempter when the idea struck me that, if I did not complete the work whilst the fit was on me, I should never complete it; so I held on. I am almost afraid to state how many pages I wrote that day of the life of Joseph Sell.

From this time I proceeded in a somewhat more leisurely manner; but, as I drew nearer to the completion of my task, dreadful fears and despondencies came over me. It will be too late, thought I; by the time I have finished the work, the bookseller will have been supplied with a tale or a novel. Is it probable that, in a town like this, where talent is so abundant — hungry talent, too — a bookseller can advertise for a tale or a novel without being supplied with half a dozen in twenty-four hours? I may as well fling down my pen — I am writing to no purpose. And these thoughts came over my mind so often, that at last, in utter despair, I flung down the pen. Whereupon the tempter within me said: 'And, now you have flung down the pen, you may as well fling yourself out of the window; what remains for you to do?' Why, to take it up again, thought I to myself, for I did not like the latter suggestion at all — and then forthwith I resumed the pen, and wrote with greater vigour than before, from about six o'clock in the evening until I could hardly see, when I rested for a while, when the tempter within me again said, or appeared to say: 'All you have been writing is stuff, it will never do — a drug — a mere drug', and methought these last words were uttered in the gruff tones of the big publisher. 'A thing merely to be sneezed at,' a voice like that of Taggart[4] added; and then I seemed to hear a sternutation — as I probably did, for, recovering from a kind of swoon, I found myself shivering with cold. The next day I brought my work to a conclusion.

But the task of revision still remained; for an hour or two I shrank from it, and remained gazing stupidly at the pile of paper which I had written over. I was all but exhausted, and I dreaded, on inspecting the sheets, to find them full of absurdities which I had paid no regard to in the furor of composition. But the task, however trying on my nerves, must be got over; at last, in a kind of desperation, I entered upon it. It was far from an easy one; there were, however, fewer errors and absurdities than I had anticipated. About twelve o'clock at night I had got over the task of revision. 'To-morrow, for the bookseller,' said I, as my head sank on the pillow. 'Oh me!' . . .

On arriving at the bookseller's shop, I cast a nervous look at the window, for the purpose of observing whether the paper had been removed or not. To my great delight the paper was in its place; with a beating heart I entered, there was nobody in the shop; as I stood at the counter, however, deliberating whether or not I should call out, the door of what seemed to be a back-parlour opened, and out came a well-dressed ladylike female, of about thirty, with a good-looking and intelligent countenance. 'What is your business, young man?' said she to me, after I had made her a polite bow. 'I wish to speak to the gentleman of the house,' said I. 'My husband is not within at present,' she replied; 'what is your business?' 'I have merely brought something to show him,' said I, 'but I will call again.' 'If you are the young gentleman who has been here before,' said the lady, 'with poems and ballads, as, indeed, I know you are,' she added, smiling, 'for I have seen you through the glass door, I am afraid it will be useless; that is,' she added with another smile, 'if you bring us nothing else.' 'I have not brought you poems and ballads now,' said I, 'but something widely different; I saw your advertisement for a tale or a novel, and have written something which I think will suit; and here it is,' I added, showing the roll of paper which I held in my hand. 'Well,' said the bookseller's wife, 'you may leave it, though I cannot promise you much chance of its being accepted. My husband has already had several offered to him; however, you may leave it; give it to me. Are you afraid to entrust it to me?' she demanded somewhat hastily, observing that I hesitated. 'Excuse me,' said I, 'but it is all I have to depend upon in the world; I am chiefly apprehensive that it will not be read.' 'On that point I can reassure you,' said the good lady, smiling, and there was now something sweet in her smile. 'I give you my word that it shall be read; come again to-morrow morning at eleven, when, if not approved, it shall be returned to you.'

I returned to my lodging, and forthwith betook myself to bed, notwithstanding the earliness of the hour. I felt tolerably tranquil; I had now cast my last stake, and was prepared to abide by the result. Whatever that result might be, I could have nothing to reproach myself with; I had strained all the energies which nature had given me in order to rescue myself from the difficulties which surrounded me. I presently sank into a sleep, which endured during the remainder of the day and the whole of the succeeding night. I awoke about nine on the morrow, and spent my last threepence on a breakfast somewhat more luxurious than the immediately preceding ones, for one penny of the sum was expended on the purchase of milk.

At the appointed hour I repaired to the house of the bookseller; the bookseller was in his shop. 'Ah,' said he, as soon as I entered, 'I am glad to see you.' There was an unwonted heartiness in the bookseller's tones, an unwonted benignity in his face. 'So,' said he, after pause, 'you have taken my advice, written a book of adventure; nothing like taking the advice, young man, of your superiors in age. Well, I think your book will do, and so does my wife, for whose judgement I have a great regard; as well I may say, as she is the daughter of a first-rate novelist, deceased. I think I shall venture on sending your book to the press.' 'But,' said I, 'we have not yet agreed upon terms.' 'Terms, terms,' said the bookseller; 'ahem! well, there is nothing like coming to terms at once. I will print the book, and give you half the profit when the edition is sold.' 'That will not do,' said I; 'I intend shortly to leave London; I must have something at once.' 'Ah, I see,' said the bookseller, 'in distress; frequently the case with authors, especially young ones. Well, I don't care if I purchase it of you, but you must be moderate; the public are very fastidious, and the speculation may prove a losing one, after all. Let me see, will five —— hem' — he stopped. I looked the bookseller in the face; there was something peculiar in it. Suddenly it appeared to me as if the voice of him of the thimble[5] sounded in my ear: 'Now is your time, ask enough, never such another chance of establishing yourself; respectable trade, pea and thimble.' 'Well,' said I at last, 'I have no objection to take the offer which you were about to make, though I really think five-and-twenty guineas to be scarcely enough, everything considered.' 'Five-and-twenty guineas!' said the bookseller; 'are you — what was I going to say — I never meant to offer half as much — I mean a quarter; I was going to say five guineas — I mean pounds; I will, however, make it up guineas.' 'That will not do,' said I; 'but, as I find we shall not deal, return me my manuscript, that I may carry it to some one else.' The bookseller looked blank. 'Dear me,' said he, 'I should never have supposed that you would have made any objection to such an offer; I am quite sure that you would have been glad to take five pounds for either of the two huge manuscripts of songs and ballads that you brought me on a former occasion.' 'Well,' said I, 'if you will engage to publish either of those two manuscripts, you shall have the present one for five pounds.' 'God forbid that I should make any such bargain,' said the bookseller; 'I would publish neither on any account; but , with respect to this last book, I have really an inclination to print it, both for your sake and mine; suppose we say ten pounds.' 'No,' said I, 'ten pounds will not do; pray restore me my manuscript.' 'Stay,' said the bookseller, 'my wife is in the next room, I will go and consult her.' Thereupon he went into his back room, where I heard him conversing with his wife in a low tone; in about ten minutes he returned. 'Young gentleman,' said he, 'perhaps you will take tea with us this evening, when we will talk further over the matter.'

That evening I went and took tea with the bookseller and his wife, both of whom, particularly the latter, overwhelmed me with civility. It was not long before I learned that the work had already been sent to the press, and was intended to stand at the head of a series of entertaining narratives, from which my friends promised themselves considerable profit. The subject of terms was again brought forward. I stood firm to my first demand for a long time; when, however, the bookseller's wife complimented me on my production in the highest terms, and said that she discovered therein the germs of genius, which she made no doubt would some day prove ornamental to my native land, I consented to drop my demand to twenty pounds, stipulating, however, that I should not be troubled with the correction of the work.

Before I departed I received the twenty pounds, and departed with a light heart to my lodgings.

Reader, amidst the difficulties and dangers of this life, should you ever be tempted to despair, call to mind these latter chapters of the life of Lavengro. There are few positions, however difficult, from which dogged resolution and perseverance may not liberate you.

1126 [1]

Shall I take thee, the Poet said
To the propounded word?
Be stationed with the Candidates
Till I have finer tried —

The Poet searched Philology
And when about to ring
For the suspended Candidate
There came unsummoned in —

That portion of the Vision
The World applied to fill
Not unto nomination
The Cherubim reveal —

1212

A word is dead
When it is said,
Some say.
I say it just
Begins to live
That day.

1263

There is no Frigate like a Book
To take us Lands away
Nor any Coursers like a Page
Of prancing Poetry —
This Traverse may the poorest take
Without oppress of Toll —
How frugal is the Chariot
That bears the Human soul.

1409

Could mortal lip divine
The undeveloped Freight
Of a delivered syllable
'Twould crumble with the weight.

1452

Your thoughts don't have words every day
They come a single time
Like signal esoteric sips
Of the communion Wine
Which while you taste so native seems
So easy so to be
You cannot comprehend its price
Nor its infrequency

Emily Dickinson (1830–86)

MARY AUGUSTA WARD

Mrs Humphry Ward, as she preferred to be called (1851–1920), granddaughter of Dr Thomas Arnold of Rugby and niece of poet Matthew Arnold, married into the intellectual and theological ferment of nineteenth-century Oxford. A prolific novelist and translator, she published *Robert Elsmere*, her best known novel, in 1888, and the autobiographical *A Writer's Recollections* in 1918.

from *A Writer's Recollections*

The Publication of 'Robert Elsmere'

It was in 1885, after the completion of the Amiel translation,[1] that I began 'Robert Elsmere,' drawing the opening scenes from that expedition to Long Sleddale in the spring of that year which I have already mentioned. The book took me three years — nearly — to write. Again and again I found myself dreaming that the end was near, and publication only a month or two away; only to sink back on the dismal conviction that the second, or the first, or the third volume — or some portion of each — must be rewritten, if I was to satisfy myself at all. I actually wrote the last words of the last chapter in March 1887, and came out afterwards, from my tiny writing-room at the end of the drawing room, shaken with tears, and wondering as I sat alone on the floor, by the fire, in the front room, what life would be like now that the book was done! But it was nearly a year after that before it came out, a year of incessant hard work, of endless re-writing, and much nervous exhaustion. For all the work was saddened and made difficult by the fact that my mother's long illness was nearing its end, and that I was torn incessantly between the claim of the book, and the desire to be with her whenever I could possibly be spared from my home and children. Whenever there was a temporary improvement in her state, I would go down to Borough alone to work feverishly at revision, only to be drawn back to her side before long by worse news. And all the time London life went on as usual, and the strain at times was great.

The difficulty of finishing the book arose first of all from its length. I well remember the depressed countenance of Mr. George Smith — who was to be to me through fourteen years afterwards the kindest of publishers and friends — when I called one day in Waterloo Place, bearing a basketful of type-written sheets. 'I am afraid you have brought us a perfectly unmanageable book!' he said; and I could only mournfully agree that so it was. It was far too long, and my heart sank at the thought of all there was still to do. But how patient Mr. Smith was over it! — and how generous in the matter of unlimited fresh proofs and endless corrections. I am certain that he had no belief in the book's success; and yet on the ground of his interest in 'Miss Bretherton'[2] he had made liberal terms with me, and all through the long incubation he was always indulgent and sympathetic.

The root difficulty was of course the dealing with such a subject in a novel at all. Yet I was determined to deal with it so, in order to reach the public. There were great precedents — Froude's 'Nemesis of Faith,' Newman's 'Loss and Gain,' Kingsley's 'Alton Locke,' — for the novel of religious or social propaganda. And it seemed to me that the novel was capable of holding and shaping real experience of any kind, as it affects the lives of men and women. It is the most elastic, the most adaptable of forms. No one has a right to set limits to its range. There is only one final test. Does it interest? — does it appeal? Personally, I should add another. Does it make in the long run for *beauty*? Beauty taken in the largest and most generous sense, and especially as including discord, the harsh and jangled notes which enrich the rest — but still Beauty — as Tolstoy was a master of it.

But at any rate, no one will deny that *interest* is the crucial matter.

> There are five and twenty ways
> Of constructing tribal lays —
> And every single one of them is right![3]

— always supposing that the way chosen quickens the breath and stirs the heart of those who listen. But when the subject chosen has two aspects, the one intellectual and logical, the other poetic and emotional, the difficulty of holding the balance between them so that neither overpowers the other, and interest is maintained, is admittedly great.

I wanted to show how a man of sensitive and noble character, born for religion, comes to throw off the orthodoxies of his day and moment, and to go out into the wilderness where all is experiment, and spiritual life begins again. And with him I wished to contrast a type no less fine of the traditional and guided mind — and to try to imagine the clash of two such tendencies of thought, as it might affect all practical life, and especially the life of two people who loved each other.

Here then — to begin with — were Robert and Catherine. Yes — but Robert must be made intellectually intelligible. Closely looked at, all novel-writing is a sort of shorthand. Even the most simple and broadly human situation cannot be told in full. Each reader in following it unconsciously supplies a vast amount himself. A great deal of the effect is owing to things quite out of the picture given — things in the reader's own mind, first and foremost. The writer is playing on common experience; and mere suggestion is often far more effective than analysis. Take the paragraph in Turguénieff's 'Lisa' — it was pointed out to me by Henry James — where Lavretsky on the point of marriage, after much suffering, with the innocent and noble girl whom he adores, suddenly hears that his intolerable first wife whom he had long believed dead is alive. Turguénieff, instead of setting out the situation in detail, throws himself on the reader. 'It was dark. Lavretsky went into the garden, and walked up and down there till dawn.'

That is all. And it is enough. The reader who is not capable of sharing that night walk with Lavretsky, and entering into his thoughts, has read the novel to no purpose. He would not understand, though Lavretsky or his creator were to spend pages on explaining.

But in my case, what provoked the human and emotional crisis — what produced the *story* — was an intellectual process. Now the difficulty here in using suggestion — which is the master tool of the novelist — is much greater than in the case of ordinary experience. For the conscious use of the intellect on the accumulated data of life — through history and philosophy — is not ordinary experience. In its more advanced forms, it only applies to a small minority of the human race.

Still, in every generation, while a minority is making or taking part in the intellectual process itself, there is an atmosphere, a diffusion, produced around them, which affects many many thousands who have but little share — but little *conscious* share, at any rate — in the actual process.

Here then is the opening for suggestion — in connection with the various forms of imagination which enter into Literature; with poetry, and fiction, which, as Goethe saw, is really a form of poetry. And a quite legitimate opening. For to use it is to quicken the intellectual process itself, and to induce a larger number of minds to take part in it.

The problem then, in intellectual poetry or fiction, is so to suggest the argument, that both the expert and the popular consciousness may feel its force. And to do this without overstepping the bounds of poetry or fiction; without turning either into mere ratiocination, and so losing the 'simple, sensuous, passionate' element which is their true life.

It was this problem which made 'Robert Elsmere' take three years to write instead of one. Mr. Gladstone complained in his famous review of it that a majestic system which had taken centuries to elaborate, and gathered into itself the wisest brains of the ages had gone down in a few weeks or months before the onslaught of the Squire's arguments; and that if the Squire's arguments were few the orthodox arguments were fewer! The answer to the first part of the charge is that the well-taught schoolboy of to-day is necessarily wiser in a hundred respects than Sophocles or Plato, since he represents not himself, but the brainwork of a hundred generations since those great men lived. And as to the second, if Mr. Gladstone had seen the first redactions of the book — only if he had, I fear he would never have read it! — he would hardly have complained of lack of argument on either side, whatever he might have thought of its quality. Again and again I went on writing for hours, satisfying the logical sense in one's self, trying to put the arguments on both sides as fairly as possible, only to feel despairingly at the end that it must all come out. It might be decent controversy; but life, feeling, charm, *humanity* had gone out of it; it had ceased therefore to be 'making,' to be literature.

So that in the long run there was no other method possible than suggestion — and, of course *selection*! — as with all the rest of one's material. That being understood, what one had to aim at was so to use suggestion as to touch the two zones of thought — that of the scholar, and that of what one may call the educated populace; who without being scholars, were yet aware, more or less clearly, of what the scholars were doing. It is from these last that 'atmosphere' and 'diffusion' come; the atmosphere and diffusion which alone make wide penetration for a book illustrating an intellectual motive possible. I had to learn that, having read a good deal, I must as far as possible wipe out the traces of reading. All that could be done was to leave a few sign-posts as firmly planted as one could, so as to recall the real journey to those who already knew it, and for the rest, to trust to the floating interest and passion surrounding a great controversy — the *second* religious battle of the nineteenth century — with which it had seemed to me both in Oxford and in London that the intellectual air was charged.

I grew very weary in the course of the long effort, and often very despairing. But there were omens of hope now and then; first, a letter from my dear eldest brother, the late W. T. Arnold, who died in 1904, leaving a record as journalist and scholar which has been admirably told by his intimate friend and colleague, Mr. — now Captain — C. E. Montague. He and I had shared many intellectual interests connected with the history of the Empire. His monograph on 'Roman Provincial Administration,' first written as an Arnold Essay, still holds the field; and in the realm of pure literature, his one-volume edition of Keats is there to show his eagerness for beauty and his love of English verse. I sent him the very first volume in proof, about a year before the book came out, and awaited his verdict with much anxiety. It came one May day in 1889. I happened to be very tired and depressed at the moment, and I remember sitting alone for a little while with the letter in my hand, without courage to open it. Then at last I opened it.

Warm congratulation — Admirable! — Full of character and colour. . . . 'Miss Bretherton' was an intellectual exercise. This is quite a different affair, and has interested and touched me deeply, as I feel sure it will all the world. The biggest thing that — with a few other things of the same kind — has been done for years.

Well! — that was enough to go on with, to carry me through the last wrestle with proofs and revision. But by the following November, nervous fatigue made me put work aside for a few weeks, and we went abroad for rest, only to be abruptly summoned home by my mother's state. Thenceforward I lived a double life — the one overshadowed by my mother's approaching death, the other amid the agitation of the book's appearance, and all the incidents of its rapid success.

I have already told the story in the Introduction to the Library Edition of 'Robert Elsmere,' and I will only run through it here, as rapidly as possible, with a few fresh incidents and quotations. There was never any doubt at all of the book's fate, and I may repeat again that before Mr. Gladstone's review of it the three volumes were already in a third edition, the rush at all the libraries was in full course, and Matthew Arnold — so gay and kind, in those March weeks before his own sudden death! — had clearly foreseen the rising boom. 'I shall take it with me to Bristol next week and get through it there, I hope [but he didn't achieve it!]. It is one of my regrets not to have known the Green of your dedication.' And a week or two later he wrote an amusing letter to his sister describing a country-house party at beautiful Wilton, Lord Pembroke's home near Salisbury, and the various stages in the book reached by the members of the party, including Mr. Goschen, who were all reading it, and all talking of it. I never, however, had any criticism of it from him, except of the first volume, which he liked. I doubt very much whether the second and third volumes would have appealed to him. My uncle was a Modernist long before the time. In 'Literature and Dogma,' he threw out in detail much of the argument suggested in 'Robert Elsmere,' but to the end of his life he was a contented member of the Anglican Church, so far as attendances at her services was concerned, and belief in her mission of 'edification' to the English people. He had little sympathy with people who 'went out.' Like Mr. Jowett, he would have liked to see the Church slowly reformed and 'modernised' from within. So that with the main theme of my book — that a priest who doubts must depart — he could never have had full sympathy. And in the course of years — as I showed in a later novel written twenty-four years after 'Robert Elsmere' — I feel that I have very much come to agree with him! These great national structures that we call churches are too precious for iconoclast handling, if any other method is possible. The strong assertion of individual liberty within them, as opposed to the attempt to break them down from without: — that seems to me now the hopeful course. A few more heresy trials like those which sprang out of 'Essays and Reviews,' or the persecution of Bishop Colenso, would let in fresh life and healing nowadays, as did those old stirrings of the waters. The first Modernist bishop who stays in his place, forms a Modernist chapter and diocese around him, and fights the fight where he stands, will do more for liberty and faith in the Church, I now sadly believe, than those scores of brave 'forgotten dead' who have gone out of her for conscience' sake, all these years.

But to return to the book. All through March the tide of success was rapidly rising; and when I was able to think of it, I was naturally carried away by the excitement and astonishment of it. But with the later days of March a veil dropped between me and the book. My mother's suffering and storm-beaten life was coming rapidly to its close, and I could think of nothing else. In an interval of slight improvement, indeed, when it seemed as though she might rally for a time, I heard Mr. Gladstone's name quoted for the first time in connection with the book. It will be remembered that he was then out of office, having been overthrown on the Home Rule Question in '86, and he happened to be staying for an Easter visit with the Warden of Keble, and Mrs. Talbot, who was his niece by marriage. I was with my mother about a mile away, and Mrs. Talbot, who came to ask for news of her, reported to me that Mr. Gladstone was deep in the book. He was reading it pencil in hand, marking all the passages he disliked or quarrelled with, with the Italian '*Ma!*' — and those he approved of with mysterious signs which she who followed him through the volume could not always decipher. Mr. Knowles, she reported, the busy editor of the *Nineteenth Century*, was trying to persuade the great man to review it. But 'Mr. G.' had not made up his mind.

Then all was shut out again. Through many days my mother asked constantly for news of the book, and smiled with a flicker of her old brightness, when anything pleased her in a letter or review. But finally there came long hours when to think or speak of it seemed sacrilege. And on April 7 she died.

To R. B.[1]

The fine delight that fathers thought; the strong
Spur, live and lancing like the blowpipe flame,
Breathes once and, quenchèd faster than it came,
Leaves yet the mind a mother of immortal song.
Nine months she then, nay years, nine years she long
Within her wears, bears, cares and combs the same:
The widow of an insight lost she lives, with aim
Now known and hand at work now never wrong.

    Sweet fire the sire of muse, my soul needs this;
I want the one rapture of an inspiration.
O then if in my lagging lines you miss
The roll, the rise, the carol, the creation,
My winter world, that scarcely breathes that bliss
Now, yields you, with some sighs, our explanation.

Gerard Manley Hopkins (1844–89)

OSCAR WILDE

Oscar Wilde (1854–1900), writing with characteristic wit and paradox, explores the role of the artist and particularly that of the writer. By 'lying' he means something closely akin to what Chaucer called 'makyng' and Sir Philip Sidney called 'poetry'. Behind Wilde, for all his Romanticism and iconoclasm, we may similarly detect the influence of Plato and Aristotle.

from *The Decay of Lying*

An observation

A Dialogue.
Persons: Cyril and Vivian.[1]
Scene: the library of a country house
in Nottinghamshire.

C yril (coming in through the open window from the terrace). My dear Vivian, don't coop yourself up all day in the library. It is a perfectly lovely afternoon. The air is exquisite. There is a mist upon the woods, like the purple bloom upon a plum. Let us go and lie on the grass, and smoke cigarettes, and enjoy nature.

Vivian. Enjoy Nature! I am glad to say that I have entirely lost that faculty. People tell us that Art makes us love Nature more than we loved her before; that it reveals her secrets to us; and that after a careful study of Corot and Constable we see things in her that had escaped our observation. My experience is that the more we study Art, the less we care for Nature. What Art really reveals to us is Nature's lack of design, her curious crudities, her extraordinary monotony, her absolutely unfinished condition. Nature had good intentions, of course, but, as Aristotle once said, she cannot carry them out. When I look at a landscape I cannot help seeing all its defects. It is fortunate for us however that Nature is so imperfect, as otherwise we should have had no art at all. Art is our spirited protest, our gallant attempt to teach Nature her proper place. As for the infinite variety of Nature, that is a pure myth. It is not to be found in Nature herself. It resides in the imagination, or fancy, or cultivated blindness of the man who looks at her.

Cyril. Well, you need not look at the landscape. You can lie on the grass and smoke and talk.

Vivian. But Nature is so uncomfortable. Grass is hard and lumpy and damp, and full of dreadful black insects. Why, even Morris' poorest workman could make you a more comfortable seat than the whole of Nature can. Nature pales before the furniture of 'the street which from Oxford has borrowed its name,' as the poet you love so much once vilely phrased it.[2] I don't complain. If Nature had been comfortable, mankind would never have invented architecture, and I prefer houses to the open air. In a house we all feel of the proper proportions. Everything is subordinated to us, fashioned for our use and our pleasure. Egotism itself, which is so necessary to a proper sense of human dignity, is entirely the result of indoor life. Out of doors one becomes abstract and impersonal. One's individuality absolutely leaves one. And then Nature is so indifferent, so unappreciative. Whenever I am walking in the park here, I always feel that I am no more to her than the cattle that browse on the slope, or the burdock that blooms in the ditch. Nothing is more evident than that Nature hates Mind. Thinking is the most unhealthy thing in the world, and people die of it just as they die of any other disease. Fortunately, in England at any rate, thought is not catching. Our splendid physique as a people is entirely due to our national stupidity. I only hope we shall be able to keep this great historic bulwark of our happiness for many years to come; but I am afraid that we are beginning to be over-educated; at least everybody who is incapable of learning has taken to teaching — that is really what our enthusiasm for education has come to. In the meantime, you had better go back to your wearisome uncomfortable Nature, and leave me to correct my proofs.

Cyril. Writing an article! That is not very consistent after what you have just said.

Vivian. Who wants to be consistent? The dullard and the doctrinaire, the tedious people who carry out their principles to the bitter end of action, to the *reductio ad absurdum* of practice. Not I. Like Emerson, I write over the door of my library the word 'Whim.' Besides, my article is really a most salutary and valuable warning. If it is attended to, there may be a new Renaissance of Art.

Cyril. What is the subject?

Vivian. I intend to call it 'The Decay of Lying: A Protest.'

Cyril. Lying! I should have thought that our politicians kept up that habit.

Vivian. I assure you that they do not. They never rise beyond the level of misrepresentation, and actually condescend to prove, to discuss, to argue. How different from the temper of the true liar, with his frank, fearless statements, his superb irresponsibility, his healthy, natural disdain of proof of any kind! After all, what is a fine lie? Simply that which is its own evidence. If a man is sufficiently unimaginative to produce evidence in support of a lie, he might just as well speak the truth at once. No, the politicians won't do. Something may, perhaps, be argued on behalf of the Bar. The mantle of the Sophist has fallen on its members. Their feigned ardours and unreal rhetoric are delightful. They can make the worse appear the better cause, as though they were fresh from Leontine[3] schools, and have been known to wrest from reluctant juries triumphant verdicts of acquittal for their clients, even when those clients, as often happens, were clearly and unmistakeably innocent. But they are briefed by the prosaic, and are not ashamed to appeal to precedent. In spite of their endeavours, the truth will out. Newspapers, even, have degenerated. They may now be absolutely relied upon. One feels it as one wades through their columns. It is always the unreadable that occurs. I am afraid that there is not much to be said in favour of either the lawyer or the journalist. Besides, what I am pleading for is Lying in art. Shall I read you what I have written? It might do you a great deal of good. . . .

Cyril. You will find me all attention.

Vivian (*reading in a very clear, musical voice*). 'THE DECAY OF LYING: A PROTEST. — One of the chief causes that can be assigned for the curiously commonplace character of most of the literature of our age is undoubtedly the decay of Lying as an art, a science, and a social pleasure. The ancient historians gave us delightful fiction in the form of fact; the modern novelist presents us with dull facts under the guise of fiction. The Blue-Book[4] is rapidly becoming his ideal both for method and manner. He has his tedious "*document humain*," his miserable little "*coin de la création*," [5] into which he peers with his microscope. He is to be found at the Librairie Nationale, or at the British Museum, shamelessly reading up his subject. He has not even the courage of other people's ideas, but insists on going directly to life for everything, and ultimately, between encyclopædias and personal experience, he comes to the ground, having drawn his types from the family circle or from the weekly washerwoman, and having acquired an amount of useful information from which never, even in his most meditative moments, can he thoroughly free himself.

'The loss that results to literature in general from this false ideal of our time can hardly be overestimated. People have a careless way of talking about a "born liar", just as they talk about a "born poet". But in both cases they are wrong. Lying and poetry are arts — arts, as Plato saw, not unconnected with each other — and they require the most careful study, the most disinterested devotion. Indeed, they have their technique, just as the more material arts of painting and sculpture have their subtle secrets of form and colour, their craft-mysteries, their deliberate artistic methods. As one knows the poet by his fine music, so one can recognize the liar by his rich rhythmic utterance, and in neither case will the casual inspiration of the moment suffice. Here, as elsewhere, practice must precede perfection. But in modern days while the fashion of writing poetry has become far too common, and should, if possible, be discouraged, the fashion of lying has almost fallen into disrepute. Many a young man starts in life with a natural gift for exaggeration which, if nurtured in congenial and sympathetic surroundings, or by the imitation of the best models, might grow into something really great and wonderful. But, as a rule, he comes to nothing. He either falls into careless habits of accuracy —— '

Cyril. My dear fellow!

Vivian. Please don't interrupt in the middle of a sentence. 'He either falls into careless habits of accuracy, or takes to frequenting the society of the aged and the well-informed. Both things are equally fatal to his imagination, as indeed they would be fatal to the imagination of anybody, and in a short time he develops a morbid and unhealthy faculty of truth-telling, begins to verify all statements made in his presence, has no hesitation in contradicting people who are much younger than himself, and often ends by writing novels which are so like life that no one can possibly believe in their probability. This is no isolated instance that we are giving. It is simply one out of many; and if something cannot be done to check, or at least modify, our monstrous worship of facts, Art will become sterile, and Beauty will pass away from the land.

'Even Mr. Robert Louis Stevenson, that delightful master of delicate and fanciful prose, is tainted with this modern vice, for we know positively no other name for it. There is such a thing as robbing a story of its reality by trying to make it too true, and *The Black Arrow* is so inartistic as not to contain a single anachronism to boast of, while the transformation of Dr. Jekyll reads dangerously like an experiment out of the *Lancet*. . . .

'Facts are not merely finding a footing-place in history, but they are usurping the domain of Fancy, and have invaded the kingdom of Romance. Their chilling touch is over everything. They are vulgarising mankind. The crude commercialism of America, its materialising spirit, its indifference to the poetical side of things, and its lack of imagination and of high unattainable ideals, are entirely due to that country having adopted for its national hero a man, who according to his own confession, was incapable of telling a lie, and it is not too much to say that the story of George Washington and the cherry-tree has done more harm, and in a shorter space of time, than any other moral tale in the whole of literature.'

Cyril. My dear boy!

Vivian. I assure you it is the case, and the amusing part of the whole thing is that the story of the cherry-tree is an absolute myth. However, you must not think that I am too despondent about the artistic future either of America or of our own country. Listen to this: —

'That some change will take place before this century has drawn to its close we have no doubt whatsoever. Bored by the tedious and improving conversation of those who have neither the wit to exaggerate nor the genius to romance, tired of the intelligent person whose reminiscences are always based upon memory, whose statements are invariably limited by probability, who is at any time liable to be corroborated by the merest Philistine who happens to be present, Society sooner or later must return to its lost leader, the cultured and fascinating liar. Who he was who first, without ever having gone out to the rude chase, told the wondering cavemen at sunset how he had dragged the Megatherium from the purple darkness of its jasper cave, or slain the Mammoth in single combat and brought back its gilded tusks, we cannot tell, and not one of our modern anthropologists, for all their much-boasted science, has had the ordinary courage to tell us. Whatever was his name or race, he certainly was the true founder of social intercourse. For the aim of the liar is simply to charm, to delight, to give pleasure. He is the very basis of civilized society, and without him a dinner party, even at the mansions of the great, is as dull as a lecture at the Royal Society, or a debate at the Incorporated Authors, or one of Mr. Burnand's farcical comedies.[6]

'Nor will he be welcomed by society alone. Art, breaking from the prison-house of realism, will run to greet him, and will kiss his false, beautiful lips, knowing that he alone is in possession of the great secret of all her manifestations, the secret that Truth is entirely and absolutely a matter of style; while Life — poor, probable, uninteresting human life — tired of repeating herself for the benefit of Mr. Herbert Spencer,[7] scientific historians, and the compilers of statistics in general, will follow meekly after him, and try to produce, in her own simple and untutored way, some of the marvels of which he talks.

'No doubt there will always be critics who, like a certain writer in the *Saturday Review*, will gravely censure the teller of fairy tales for his defective knowledge of natural history, who will measure imaginative work by their own lack of any imaginative faculty, and will hold up their inkstained hands in horror if some honest gentleman, who has never been farther than the yew-trees of his own garden, pens a fascinating book of travels like Sir John Mandeville, or, like great Raleigh, writes a whole history of the world, without knowing anything whatsoever about the past. To excuse themselves they will try and shelter under the shield of him who made Prospero the magician, and gave him Caliban and Ariel as his servants, who heard the Tritons blowing their horns round the coral reefs of the Enchanted Isle, and the fairies singing to each other in a wood near Athens, who led the phantom kings in dim procession across the misty Scottish heath, and hid Hecate in a cave with the weird sisters.[8] They will call upon Shakespeare — they always do — and will quote that hackneyed passage about Art holding the mirror up to Nature, forgetting that this unfortunate aphorism is deliberately said by Hamlet in order to convince the bystanders of his absolute insanity in all art-matters.' [9]

Cyril. Ahem! Another cigarette, please.

Vivian. My dear fellow, whatever you may say, it is merely a dramatic utterance, and no more represents Shakespeare's real views upon art than the speeches of Iago represent his real views upon morals.[10] But let me get to the end of the passage:

'Art finds her own perfection within, and not outside of, herself. She is not to be judged by any external standard of resemblance. She is a veil, rather than a mirror. She has flowers that no forests know of, birds that no woodland possesses. She makes and unmakes many worlds, and can draw the moon from heaven with a scarlet thread. Hers are the "forms more real than living man,"[11] and hers the great archetypes of which things that have existence are but unfinished copies. Nature has, in her eyes, no laws, no uniformity. She can work miracles at her will, and when she calls monsters from the deep they come.[12] She can bid the almond tree blossom in winter, and send the snow upon the ripe cornfield. At her word the frost lays its silver finger on the burning mouth of June, and the winged lions creep out from the hollows of the Lydian hills. The dryads peer from the thicket as she passes by, and the brown fauns smile strangely at her when she comes near them. She has hawk-faced gods that worship her, and the centaurs gallop at her side.'

Cyril. I like that. I can see it. Is that the end?

Vivian. No. There is one more passage, but it is purely practical. It simply suggests some methods by which we could revive this lost art of Lying.

Cyril. Well, before you read it to me, I should like to ask you a question. What do you mean by saying that life, 'poor, probable, uninteresting human life,' will try to reproduce the marvels of art? I can quite understand your objection to art being treated as a mirror. You think it would reduce genius to the position of a cracked looking-glass. But you don't mean to say that you seriously believe that Life imitates Art, that Life in fact is the mirror, and Art the reality?

Vivian. Certainly I do. Paradox though it may seem — and paradoxes are always dangerous things — it is none the less true that Life imitates art far more than Art imitates life. . . .

The most obvious and the vulgarest form in which this is shown is in the case of the silly boys who, after reading the adventures of Jack Sheppard or Dick Turpin,[13] pillage the stalls of unfortunate apple-women, break into sweet-shops at night, and alarm old gentlemen who are returning home from the city by leaping out on them in suburban lanes, with black masks and unloaded revolvers. This interesting phenomenon, which always occurs after the appearance of a new edition of either of the books I have alluded to, is usually attributed to the influence of literature on the imagination. But this is a mistake. The imagination is essentially creative and always seeks for a new form. The boy-burglar is simply the inevitable result of life's imitative instinct. He is Fact, occupied as Fact usually is, with trying to reproduce Fiction, and what we see in him is repeated on an extended scale throughout the whole of life. Schopenhauer[14] has analysed the pessimism that characterises modern thought, but Hamlet invented it. The world has become sad because a puppet was once melancholy. The Nihilist, that strange martyr who has no faith, who goes to the stake without enthusiasm, and dies for what he does not believe in, is a purely literary product.[15] He was invented by Tourgénieff, and completed by Dostoieffski. Robespierre came out of the pages of Rousseau as surely as the People's Palace rose out of the *débris* of a novel.[16] Literature always anticipates life. It does not copy it, but moulds it to its purpose. . . .

However, I do not wish to dwell any further upon individual instances. Personal experience is a most vicious and limited circle. All that I desire to point out is the general principle that Life imitates Art far more than Art imitates Life, and I feel sure that if you think seriously about it you will find that it is true. Life holds the mirror up to Art and either reproduces some strange type imagined by painter or sculptor, or realizes in fact what has been dreamed in fiction. Scientifically speaking, the basis of life — the energy of life, as Aristotle would call it — is simply the desire for expression, and Art is always presenting various forms through which this expression can be attained. Life seizes on them and uses them, even if they be to her own hurt. Young men have committed suicide because Rolla did so, have died by their own hand because by his own hand Werther died.[17] Think of what we owe to the imitation of Christ, of what we owe to the imitation of Caesar.

Cyril. The theory is certainly a very curious one, but to make it complete you must show that Nature, no less than Life, is an imitation of Art. Are you prepared to prove that?

Vivian. My dear fellow, I am prepared to prove anything. . . .

However, I must read the end of my article: —

'What we have to do, what at any rate it is our duty to do, is to revive this old art of Lying. Much of course may be done, in the way of educating the public, by amateurs in the domestic circle, at literary lunches, and at afternoon teas. But this is merely the light and graceful side of lying, such as was probably heard at Cretan dinner parties. There are many other forms. Lying for the sake of gaining some immediate personal advantage, for instance — lying with a moral purpose, as it is usually called — though of late it has been rather looked down upon, was extremely popular with the antique world. Athena laughs when Odysseus tells her "his words of sly devising," [18] as Mr. William Morris phrases it, and the glory of mendacity illumines the pale brow of the stainless hero of Euripidean tragedy, and sets among the noble women of the past the young bride of one of Horace's most exquisite odes.[19] Later on, what at first had been merely a natural instinct was elevated into a self-conscious science. Elaborate rules were laid down for the guidance of mankind, and an important school of literature grew up round the subject. Indeed, when one remembers the excellent philosophical treaties of Sanchez[20] on the whole question, one cannot help regretting that no one had ever thought of publishing a cheap and condensed edition of the works of that great casuist. A short primer, "When to Lie and How," if brought out in an attractive and not too expensive a form, would no doubt command a large sale, and would prove of real practical service to many earnest and deep-thinking people. Lying for the sake of the improvement of the young, which is the basis of home education, still lingers amongst us, and its advantages are so admirably set forth in the early books of Plato's *Republic* that it is unnecessary to dwell upon them here. It is a mode of lying for which all good mothers have peculiar capabilities, but it is capable of still further development, and has been sadly overlooked by the School Board. Lying for the sake of a monthly salary is of course well known in Fleet Street, and the profession of a political leader-writer is not without its advantages. But it is said to be a somewhat dull occupation, and it certainly does not lead to much beyond a kind of ostentatious obscurity. The only form of lying that is absolutely beyond reproach is Lying for its own sake, and the highest development of this is, as we have already pointed out, Lying in Art. Just as those who do not love Plato more than Truth cannot pass beyond the threshold of the Academe,[21] so those who do not love Beauty more than Truth never know the inmost shrine of Art. The solid stolid British intellect lies in the desert sands like the Sphinx in Flaubert's marvellous tale,[22] and fantasy, *La Chimère*, dances round it, and calls to it with her false, flute-toned voice. It may not hear her now, but surely some day, when we are all bored to death with the commonplace character of modern fiction, it will hearken to her and try to borrow her wings.

'And when that day dawns, or sunset reddens how joyous we shall all be! Facts will be regarded as discreditable, Truth will be found mourning over her fetters, and Romance, with her temper of wonder, will return to the land. The very aspect of the world will change to our startled eyes. Out of the sea will rise Behemoth and Leviathan,[23] and sail round the high-pooped galleys, as they do on the delightful maps of those ages when books on geography were actually readable. Dragons will wander about the waste places and the phœnix will soar from her nest of fire into the air. We shall lay our hands upon the basilisk, and see the jewel in the toad's head. Champing his gilded oats, the Hippogriff will stand in our stalls, and over our heads will float the Blue Bird singing of beautiful and impossible things, of things that are lovely and that never happen, of things that are not and that should be. But before this comes to pass we must cultivate the lost art of Lying.'

Cyril. Then we must certainly cultivate it at once. But in order to avoid making any error I want you to tell me briefly the doctrines of the new æsthetics.

Vivian. Briefly, then, they are these. Art never expresses anything but itself. It has an independent life, just as Thought has, and develops purely on its own lines. It is not necessarily realistic in an age of realism, nor spiritual in an age of faith. So far from being the creation of its time, it is usually in direct opposition to it, and the only history that it preserves for us is the history of its own progress. Sometimes it returns upon its footsteps, and revives some antique form, as happened in the archaistic movement of late Greek Art, and in the pre-Raphaelite movement of our own day. At other times it entirely anticipates its age, and produces in one century work that it takes another century to understand, to appreciate, and to enjoy. In no case does it reproduce its age. To pass from the art of a time to the time itself is the great mistake that all historians commit.

The second doctrine is this. All bad art comes from returning to Life and Nature, and elevating them into ideals. Life and Nature may sometimes be used as part of Art's rough material, but before they are of any real service to art they must be translated into artistic conventions. The moment Art surrenders its imaginative medium it surrenders everything. As a method Realism is a complete failure, and the two things that every artist should avoid are modernity of form and modernity of subject-matter. To us, who live in the nineteenth century, any century is a suitable subject for art except our own. The only beautiful things are the things that do not concern us. It is, to have the pleasure of quoting myself, exactly because Hecuba is nothing to us that her sorrows are so suitable a motive for a tragedy. Besides, it is only the modern that ever becomes old-fashioned. M. Zola sits down to give us a picture of the Second Empire. Who cares for the Second Empire[24] now? It is out of date. Life goes faster than Realism, but Romanticism is always in front of Life.

The third doctrine is that Life imitates Art far more than Art imitates Life. This results not merely from Life's imitative instinct, but from the fact that the self-conscious aim of Life is to find expression, and that Art offers it certain beautiful forms through which it may realize that energy. It is a theory that has never been put forward before, but it is extremely fruitful, and throws an entirely new light upon the history of Art.

It follows, as a corollary from this, that external Nature also imitates Art. The only effects that she can show us are effects that we have already seen through poetry, or in paintings. This is the secret of Nature's charm, as well as the explanation of Nature's weakness.

The final revelation is that Lying, the telling of beautiful untrue things, is the proper aim of Art. But of this I think I have spoken at sufficient length. And now let us go out on the terrace, where 'droops the milk-white peacock like a ghost,' while the evening star 'washes the dusk with silver.'[25] At twilight nature becomes a wonderfully suggestive effect, and is not without loveliness, though perhaps its chief use is to illustrate quotations from the poets. Come! We have talked long enough.

A Legend of Truth

Once on a time, the ancient legends tell,
Truth, rising from the bottom of her well,
Looked on the world, but, hearing how it lied,
Returned to her seclusion horrified.
There she abode, so conscious of her worth,
Not even Pilate's Question[1] called her forth,
Nor Galileo, kneeling to deny
The Laws that hold our Planet 'neath the sky.
Meantime, her kindlier sister, whom men call
Fiction, did all her work and more than all,
With so much zeal, devotion, tact, and care,
That no one noticed Truth was otherwhere.

Then came a War when, bombed and gassed and mined,
Truth rose once more, perforce, to meet mankind,
And through the dust and glare and wreck of things,
Beheld a phantom on unbalanced wings,
Reeling and groping, dazed, dishevelled, dumb,
But semaphoring direr deeds to come.
Truth hailed and bade her stand; the quavering shade
Clung to her knees and babbled, 'Sister, aid!
I am — I was — thy Deputy, and men
Besought me for my useful tongue or pen
To gloss their gentle deeds, and I complied,
And they, and thy demands, were satisfied.
But this —' she pointed o'er the blistered plain,
Where men as Gods and devils wrought amain —
'This is beyond me! Take thy work again.'

Tables and pen transferred, she fled afar,
And Truth assumed the record of the War . . .
She saw, she heard, she read, she tried to tell
Facts beyond precedent and parallel —
Unfit to hint or breathe, much less to write,
But happening every minute, day and night.
She called for proof. It came. The dossiers grew.
She marked them, first, 'Return. This can't be true.'
Then, underneath the cold official word:
'This is not really half of what occurred.'

She faced herself at last, the story runs,
And telegraphed her sister: 'Come at once.
Facts out of hand. Unable overtake
Without your aid. Come back for Truth's own sake!
Co-equal rank and powers if you agree.
They need us both, but you far more than me!'

Rudyard Kipling (1865–1936)

RUDYARD KIPLING

Rudyard Kipling (1865–1936), journalist, poet, novelist, and short-story writer, perhaps the most widely read English man of letters of his time, won the Nobel Prize for Literature in 1907. Much of his reputation rests on his writings about India, where he was born but which he left for good in his early twenties. The largely autobiographical *Something of Myself*, published in 1937, mentions an unhappy episode in his childhood: 'If you cross-examine a child of seven or eight on his day's doings (especially when he wants to go to sleep) he will contradict himself very satisfactorily. If each contradiction be set down as a lie and retailed at breakfast, life is not easy. I have known a certain amount of bullying, but this was calculated torture — religious as well as scientific. Yet it made me give attention to the lies I soon found it necessary to tell: and this, I presume, is the foundation of literary effort.' 'Working Tools', the final chapter of the book, combines anecdote with advice and information about writing.

from *Something of Myself*

Working Tools

I have told what my early surroundings were, and how richly they furnished me with material. Also, how rigorously newspaper spaces limited my canvases and, for the reader's sake, prescribed that within these limits must be some sort of beginning, middle, and end. My ordinary reporting, leader- and note-writing carried the same lesson, which took me an impatient while to learn. Added to this, I was almost nightly responsible for my output to visible and often brutally voluble critics at the Club. They were not concerned with my dreams. They wanted accuracy and interest, but first of all accuracy.

My young head was in a ferment of new things seen and realised at every turn and — that I might in any way keep abreast of the flood — it was necessary that every word should tell, carry, weigh, taste and, if need were, smell. Here the Father helped me incomparably by his 'judicious leaving alone.' 'Make your own experiments,' said he. 'It's the only road. If I helped, I'd hinder.' So I made my own experiments and, of course, the viler they were the more I admired them.

Mercifully, the mere act of writing was, and has always been, a physical pleasure to me. This made it easier to throw away anything that did not turn out well: and to practise, as it were, scales.

Verse, naturally, came first, and here the Mother was at hand, with now and then some shrivelling comment that infuriated me. But, as she said: 'There's no Mother in Poetry, my dear.' It was she, indeed, who had collected and privately printed verses written at school up to my sixteenth year, which I faithfully sent out from the little House of the Dear Ladies. Later, when the notoriety came, 'in they broke, those people of importance,' and the innocent thing 'came on to the market,' and Philadelphia lawyers, a breed by itself, wanted to know, because they had paid so much money for an old copy, what I remembered about its genesis. They had been first written in a stiff, marble-backed MS. book, the front page of which the Father had inset with a scandalous sepia-sketch of Tennyson and Browning in procession, and a spectacled schoolboy bringing up the rear. I gave it, when I left school, to a woman who returned it to me many years later — for which she will take an even higher place in Heaven than her natural goodness ensures — and I burnt it, lest it should fall into the hands of 'lesser breeds without the (Copyright) law.' [1]

I forget who started the notion of my writing a series of Anglo-Indian tales, but I remember our council over the naming of the series. They were originally much longer than when they appeared, but the shortening of them, first to my own fancy after rapturous re-readings, and next to the space available, taught me that a tale from which pieces have been raked out is like a fire that has been poked. One does not know that the operation has been performed, but every one feels the effect. Note, though, that the excised stuff must have been honestly written for inclusion. I found that when, to save trouble, I 'wrote short' *ab initio* much salt went out of the work. This supports the theory of the chimaera which, having bombinated and been removed, *is* capable of producing secondary causes *in vacuo*.

This leads me to the Higher Editing. Take of well-ground Indian ink as much as suffices and a camel-hair brush proportionate to the interspaces of your lines. In an auspicious hour, read your final draft and consider faithfully every paragraph, sentence and word, blacking out where requisite. Let it lie by to drain as long as possible. At the end of that time, re-read and you should find that it will bear a second shortening. Finally, read it aloud alone and at leisure. Maybe a shade more brushwork will then indicate or impose itself. If not, praise Allah and let it go, and 'when thou hast done, repent not.' The shorter the tale, the longer the brushwork and, normally, the shorter the lie-by, and *vice-versa*. The longer the tale, the less brush but the longer lie-by. I have had tales by me for three or five years which shortened themselves almost yearly. The magic lies in the Brush and the Ink. For the Pen, when it is writing, can only scratch; and bottled ink is not to compare with the ground Chinese stick. *Experto crede.* [2]

Let us now consider the Personal Daemon of Aristotle and others, of whom it has truthfully been written, though not published: —

This is the doom of the Makers — their Daemon lives in their pen.
If he be absent or sleeping, they are even as other men.
But if he be utterly present, and they swerve not from his behest,
The word that he gives shall continue, whether in earnest or jest.

Most men, and some most unlikely, keep him under an alias which varies
with their literary or scientific attainments. Mine came to me early when I
sat bewildered among other notions, and said: 'Take this and no other.' I
obeyed, and was rewarded. It was a tale in the little Christmas magazine
Quartette which we four wrote together, and it was called 'The Phantom
'Rickshaw.' Some of it was weak, much was bad and out of key; but it was
my first serious attempt to think in another man's skin.

After that I learned to lean upon him and recognise the sign of his
approach. If I ever held back, Ananias-fashion,[3] anything of myself (even
though I had to throw it out afterwards) I paid for it by missing what I *then*
knew the tale lacked. As an instance, many years later I wrote about a
mediaeval artist, a monastery, and the premature discovery of the microscope.
('The Eye of Allah.') Again and again it went dead under my hand, and for
the life of me I could not see why. I put it away and waited. Then said my
Daemon — and I was meditating something else at the time — 'Treat it as
an illuminated manuscript.' I had ridden off on hard black-and-white
decoration, instead of pumicing the whole thing ivory-smooth, and loading
it with thick colour and gilt. Again, in a South African, post-Boer War tale
called 'The Captive,' which was built up round the phrase 'a first-class dress-
parade for Armageddon,' I could not get my lighting into key with the tone
of the monologue. The background insisted too much. My Daemon said at
last: 'Paint the background first once for all, as hard as a public house sign,
and leave it alone.' This done, the rest fell into place with the American
accent and outlook of the teller.

My Daemon was with me in the *Jungle Books*, *Kim*, and both Puck books,
and good care I took to walk delicately, lest he should withdraw. I know that
he did not, because when those books were finished they said so themselves
with, almost, the water-hammer click of a tap turned off. One of the clauses
in our contract was that I should never follow up 'a success,' for by this sin
fell Napoleon and a few others. *Note here.* When your Daemon is in charge,
do not try to think consciously. Drift, wait, and obey.

I am afraid that I was not much impressed by reviews. But my early days
in London were unfortunate. As I got to know literary circles and their
critical output, I was struck by the slenderness of some of the writers'
equipment. I could not see how they got along with so casual a knowledge
of French work and, apparently, of much English grounding that I had
supposed indispensable. Their stuff seemed to be a day-to-day traffic in
generalities, hedged by trade considerations. Here I expect I was wrong,
but, making my own tests (the man who had asked me out to dinner to

discover what I had read gave me the notion), I would ask simple questions, misquote or misattribute my quotations; or (once or twice) invent an author. The result did not increase my reverence. Had they been newspaper men in a hurry, I should have understood; but the gentlemen were presented to me as Priests and Pontiffs. And the generality of them seemed to have followed other trades — in banks or offices — before coming to the Ink; whereas I was free born.[4] It was pure snobism on my part, but it served to keep me inside myself, which is what snobbery is for.

I would not to-day recommend any writer to concern himself overly with reviews. London is a parish, and the Provincial Press has been syndicated, standardised, and smarmed down out of individuality. But there remains still a little fun in that fair. In Manchester was a paper called *The Manchester Guardian*. Outside the mule-lines I had never met anything that could kick or squeal so continuously, or so completely round the entire compass of things. It suspected me from the first, and when my 'Imperialistic' iniquities were established after the Boer War, it used each new book of mine for a shrill recount of my previous sins (exactly as C —— used to do)[5] and, I think, enjoyed itself. In return I collected and filed its more acid but uncommonly well-written leaders for my own purposes. After many years, I wrote a tale ('The Wish House') about a woman of what was called 'temperament' who loved a man and who also suffered from a cancer on her leg — the exact situation carefully specified. The review came to me with a gibe on the margin from a faithful friend: 'You threw up a catch *that* time!' The review said that I had revived Chaucer's Wife of Bath even to the 'mormal on her shinne.' And it looked just like that too! There was no possible answer, so, breaking my rule not to have commerce with any paper, I wrote to the *Manchester Guardian* and gave myself 'out — caught to leg.' The reply came from an evident human being (I had thought red-hot linotypes composed their staff) who was pleased with the tribute to his knowledge of Chaucer.[6]

Per contra, I have had miraculous escapes in technical matters, which make me blush still. Luckily the men of the seas and the engine-room do not write to the Press, and my worst slip is still underided.

The nearest shave that ever missed me was averted by my Daemon. I was at the moment in Canada, where a young Englishman gave me, as a personal experience, a story of a body-snatching episode in deep snow, perpetrated in some lonely prairie-town and culminating in purest horror. To get it out of the system I wrote it detailedly, and it came away just a shade too good; too well-balanced; too slick. I put it aside, not that I was actively uneasy about it, but I wanted to make sure. Months passed, and I started a tooth which I took to the dentist in the little American town near 'Naulakha.' I had to wait a while in his parlour, where I found a file of bound *Harper's Magazines* — say six hundred pages to the volume — dating from the 'fifties.

I picked up one, and read as undistractedly as the tooth permitted. There I found my tale, identical in every mark — frozen ground, frozen corpse stiff in its fur robes in the buggy — the inn-keeper offering it a drink — and so on to the ghastly end. Had I published that tale, what could have saved me from the charge of deliberate plagiarism? *Note here.* Always, in our trade, look a gift horse at both ends and in the middle. He may throw you.

But here is a curious case. In the late summer, I think, of '13, I was invited to Manœuvres round Frensham Ponds at Aldershot. The troops were from the Eighth Division of the coming year — Guardsmen, Black Watch, and the rest, down to the horsed maxims — two per battalion. Many of the officers had been juniors in the Boer War, known to Gwynne, one of the guests, and some to me. When the sham fight was developing, the day turned blue-hazy, the sky lowered, and the heat struck like the Karroo, as one scuttled among the heaths, listening to the uncontrolled clang of the musketry fire. It came over me that anything might be afoot in such weather, pom-poms for instance, half heard on a flank, or the glint of a helio through a cloud-drift. In short I conceived the whole pressure of our dead of the Boer War flickering and re-forming as the horizon flickered in the heat; the galloping feet of a single horse, and a voice once well-known that passed chanting ribaldry along the flank of a crack battalion. ('But Winnie is one of the lost — poor dear!' was that song, if any remember it or its Singer in 1900–1901.) In an interval, while we lay on the grass, I told Gwynne what was in my head; and some officers also listened. The finale was to be manœuvres abandoned and a hurried calling-off of all arms by badly frightened Commandants — the men themselves sweating with terror though they knew not why.

Gwynne played with the notion, and added details of Boer fighting that I did not know; and I remember a young Duke of Northumberland, since dead, who was interested. The notion so obsessed me that I wrote out the beginning at once. But in cold blood it seemed more and more fantastic and absurd, unnecessary and hysterical. Yet, three or four times I took it up and, as many, laid it down. After the War I threw the draft away. It would have done me no good, and might have opened the door, and my mail, to unprofitable discussion. For there is a type of mind that dives after what it calls 'psychical experiences.' And I am in no way 'psychic.' Dealing as I have done with large, superficial areas of incident and occasion, one is bound to make a few lucky hits or happy deductions. But there is no need to drag in the 'clairvoyance,' or the rest of the modern jargon. I have seen too much evil and sorrow and wreck of good minds on the road to Endor[7] to take one step along that perilous track. Once only was I sure that I had 'passed beyond the bounds of ordinance.' I dreamt that I stood, in my best clothes, which I do not wear as a rule, one in a line of similarly habited men, in some vast hall, floored with rough-jointed stone slabs. Opposite me, the width of the hall, was another line of persons and the impression of a crowd behind them. On my left some ceremony was taking place that I wanted to see, but could not unless I stepped out of my line because the fat stomach of my neighbour on my left barred my vision. At the ceremony's close, both lines of spectators broke up and moved forward and met, and the great space filled with people. Then a man came up behind me, slipped his hand beneath my arm, and said: 'I want a word with you.' I forget the rest: but it had been a perfectly clear dream, and it stuck in my memory. Six weeks or more later, I attended in my capacity of a Member of the War Graves Commission a ceremony at Westminster Abbey, where the Prince of Wales dedicated a plaque to 'The Million Dead' of the Great War. We Commissioners lined up facing across the width of the Abbey Nave, more members of the Ministry and a big body of the public behind them, all in black clothes. I could see nothing of the ceremony because the stomach of the man on my left barred my vision. Then, my eye was caught by the cracks of the stone flooring and I said to myself: 'But here is where I have been!' We broke up, both lines flowed forward and met, and the Nave filled with a crowd, through which a man came up and slipped his hand upon my arm saying: 'I want a word with you, please.' It was about some utterly trivial matter that I have forgotten.

But how, and why, had I been shown an unreleased roll of my life-film? For the sake of the 'weaker brethren' — and sisters — I made no use of the experience.

In respect to verifying one's references, which is a matter in which one can help one's Daemon, it is curious how loath a man is to take his own medicine. Once, on a Boxing Day, with hard frost coming greasily out of the ground, my friend, Sir John Bland-Sutton, the head of the College of Surgeons, came down to 'Bateman's' very full of a lecture which he was to deliver on 'gizzards.' We were settled before the fire after lunch, when he volunteered that So-and-so had said that if you hold a hen to your ear, you can hear the click in its gizzard of the little pebbles that help its digestion. 'Interesting,' said I. 'He's an authority.' 'Oh yes, but' — a long pause — 'have you any hens about here, Kipling?' I owned that I had, two hundred yards down a lane, but why not accept So-and-so? 'I can't,' said John simply, 'till I've tried it.' Remorselessly, he worried me into taking him to the hens, who lived in an open shed in front of the gardener's cottage. As we skated over the glairy ground, I saw an eye at the corner of the drawn-down Boxing-Day blind, and knew that my character for sobriety would be blasted all over the farms before night-fall. We caught an outraged pullet. John soothed her for a while (he said her pulse was a hundred and twenty-six), and held her to his ear. 'She clicks all right,' he announced. 'Listen.' I did, and there was click enough for a lecture, '*Now* we can go back to the house,' I pleaded. 'Wait a bit. Let's catch that cock. He'll click better.' We caught him after a loud and long chase, and he clicked like a solitaire-board. I went home, my ears alive with parasites, so wrapped up in my own indignation that the fun of it escaped me. It had not been *my* verification, you see.

But John was right. Take nothing for granted if you can check it. Even though that seem waste-work, and has nothing to do with the essentials of things, it encourages the Daemon. There are always men who by trade or calling know the fact or the inference that you put forth. If you are wrong by a hair in this, they argue: 'False in one thing, false in all.' Having sinned, I know. Likewise, never play down to your public — not because some of them do not deserve it, but because it is bad for your hand. All your material is drawn from the lives of men. Remember, then, what David did with the water brought to him in the heat of battle.[8]

And, if it be in your power, bear serenely with imitators. My *Jungle Books* begat Zoos of them. But the genius of all the genii was one who wrote a series called *Tarzan of the Apes*. I read it, but regret I never saw it on the films, where it rages most successfully. He had 'jazzed' the motif of the *Jungle Books* and, I imagine, had thoroughly enjoyed himself. He was reported to have said that he wanted to find out how bad a book he could write and 'get away with,' which is a legitimate ambition. . . .

In the come-and-go of family talk there was often discussion as to whether I could write a 'real novel.' The Father thought that the setting of my work and life would be against it, and Time justified him.

Now here is a curious thing. At the Paris Exhibition of 1878 I saw, and never forgot, a picture of the death of Manon Lescaut, and asked my Father many questions. I read that amazing 'one book' of the Abbé Prévost, in alternate slabs with Scarron's *Roman Comique*, when I was about eighteen, and it brought up the picture.[9] My theory is that a germ lay dormant till my change of life to London (though that is not Paris) woke it up, and that *The Light that Failed* was a sort of inverted, metagrobolised phantasmagoria based on *Manon*. I was confirmed in my belief when the French took to that *conte* with relish, and I always fancied that it walked better in translation than in the original. But it was only a *conte* — not a built book.

Kim, of course, was nakedly picaresque and plotless — a thing imposed from without.

Yet I dreamed for many years of building a veritable three-decker out of chosen and long-stored timber — teak, greenheart, and ten-year-old oak knees — each curve melting deliciously into the next that the sea might nowhere meet resistance or weakness; the whole suggesting motion even when, her great sails for the moment furled, she lay in some needed haven — a vessel ballasted on ingots of pure research and knowledge, roomy, fitted with delicate cabinet-work below-decks, painted, carved, gilt and wreathed the length of her, from her blazing stern-galleries outlined by bronzy palm-trunks, to her rampant figure-head — an East Indiaman worthy to lie alongside *The Cloister and the Hearth*.

Not being able to do this, I dismissed the ambition as 'beneath the thinking mind.' So does a half-blind man dismiss shooting and golf.

Nor did I live to see the day when the new three-deckers should hoist themselves over the horizon, quivering to their own power, overloaded with bars, ball-rooms, and insistent chromium plumbing; hellishly noisy from the sports' deck to the barber's shop; but serving their generation as the old craft served theirs. The young men were already laying down the lines of them, fondly believing that the old laws of design and construction were for them abrogated.

And with what tools did I work in my own mould-loft? I had always been choice, not to say coquettish in this respect. In Lahore for my *Plain Tales* I used a slim, octagonal-sided, agate penholder with a Waverley nib. It was a gift, and when in an evil hour it snapped I was much disturbed. Then followed a procession of impersonal hirelings each with a Waverley, and next a silver penholder with a quill-like curve, which promised well but did not perform. In Villiers Street I got me an outsize office pewter ink-pot, on which I would gouge the names of the tales and books I wrote out of it. But the housemaids of married life polished those titles away till they grew as faded as a palimpsest.

I then abandoned hand-dipped Waverleys — a nib I never changed — and for years wallowed in the pin-pointed 'stylo' and its successor the 'fountain' which for me meant geyser-pens. In later years I clung to a slim, smooth, black treasure (Jael was her office name) which I picked up in Jerusalem. I tried pump-pens with glass insides, but they were of 'intolerable entrails.'

For my ink I demanded the blackest, and had I been in my Father's house, as once I was, would have kept an ink-boy to grind me Indian ink. All 'blue-blacks' were an abomination to my Daemon, and I never found a bottled vermilion fit to rubricate initials when one hung in the wind waiting.

My writing-blocks were built for me to an unchanged pattern of large, off-white, blue sheets, of which I was most wasteful. All this old-maiderie did not prevent me when abroad from buying and using blocks, and tackle, in any country.

With a lead pencil I ceased to express — probably because I had to use a pencil in reporting. I took very few notes excepts of names, dates, and addresses. If a thing didn't stay in my memory, I argued it was hardly worth writing out. But each man has his own method. I rudely drew what I wanted to remember.

Like most men who ply one trade in one place for any while, I always kept certain gadgets on my worktable, which was ten feet long from North to South and badly congested. One was a long, lacquer, canoe-shaped pen-tray full of brushes and dead 'fountains'; a wooden box held clips and bands; another, a tin one, pins; yet another, a bottle-slider, kept all manner of unneeded essentials from emery-paper to small screw-drivers; a paper-weight, said to have been Warren Hastings'; a tiny, weighted fur-seal and a leather crocodile sat on some of the papers; an inky foot-rule and a Father of Penwipers which a much-loved house-maid of ours presented yearly, made up the mainguard of these little fetishes.

My treatment of books, which I looked upon as tools of my trade, was popularly regarded as barbarian. Yet I economised on my multitudinous penknives, and it did no harm to my fore-finger. There were books I respected, because they were put in locked cases. The others, all the house over, took their chances.

Left and right of the table were two big globes, on one of which a great airman had once outlined in white paint those air-routes to the East and Australia which were well in use before my death.

A Considerable Speck

(*Microscopic*)

A speck that would have been beneath my sight
On any but a paper sheet so white
Set off across what I had written there.
And I had idly poised my pen in air
To stop it with a period of ink
When something strange about it made me think.
This was no dust speck by my breathing blown,
But unmistakably a living mite
With inclinations it could call its own.
It paused as with suspicion of my pen,
And then came racing wildly on again
To where my manuscript was not yet dry;
Then paused again and either drank or smelt —
With loathing, for again it turned to fly.
Plainly with an intelligence I dealt.
It seemed too tiny to have room for feet,
Yet must have had a set of them complete
To express how much it didn't want to die.
It ran with terror and with cunning crept.
It faltered: I could see it hesitate;
Then in the middle of the open sheet
Cower down in desperation to accept
Whatever I accorded it of fate.
I have none of the tenderer-than-thou
Collectivistic regimenting love
With which the modern world is being swept.
But this poor microscopic item now!
Since it was nothing I knew evil of
I let it lie there till I hope it slept.
I have a mind myself and recognize
Mind when I meet with it in any guise.
No one can know how glad I am to find
On any sheet the least display of mind.

Robert Frost (1874–1963)

ROBERT FROST

Robert Frost (1874–1963) celebrated his ancestral New England in poetry of universal and lasting significance. This lecture, delivered to the (male) student body of Amherst College, Massachusetts in 1930, affirms, with quiet irony, the centrality of metaphor in our understanding of the world, and hence of poetry in our educational curriculum.

Education by Poetry: A Meditative Monologue

I am going to urge nothing in my talk. I am not an advocate. I am going to consider a matter, and commit a description. And I am going to describe other colleges than Amherst. Or, rather say all that is good can be taken as about Amherst; all that is bad will be about other colleges.

I know whole colleges where all American poetry is barred — whole colleges. I know whole colleges where all contemporary poetry is barred.

I once heard of a minister who turned his daughter — his poetry-writing daughter — out on the street to earn a living, because he said there should be no more books written; God wrote one book, and that was enough. (My friend George Russell, 'Æ', has read no literature, he protests, since just before Chaucer.)

That all seems sufficiently safe, and you can say one thing for it. It takes the onus off the poetry of having to be used to teach children anything. It comes pretty hard on poetry, I sometimes think, what it has to bear in the teaching process.

Then I know whole colleges where, though they let in older poetry, they manage to bar all that is poetical in it by treating it as something other than poetry. It is not so hard to do that. Their reason I have often hunted for. It may be that these people act from a kind of modesty. Who are professors that they should attempt to deal with a thing as high and as fine as poetry? Who are *they*? There is a certain manly modesty in that.

That is the best general way of settling the problem; treat all poetry as if it were something else than poetry, as if it were syntax, language, science. Then you can even come down into the American and into the contemporary without any special risk.

There is another reason they have, and that is that they are, first and foremost in life, markers. They have the marking problem to consider. Now, I stand here a teacher of many years' experience and I have never complained of having had to mark. I had rather mark anyone for anything — for his looks, carriage, his ideas, his correctness, his exactness, anything you please — I would rather give him a mark in term of letters, A, B, C, D, than have to use adjectives on him. We are all being marked by each other all the time, classified, ranked, put in our place, and I see no escape from that. I am no sentimentalist. You have got to mark, and you have got to mark, first of all, for accuracy, for correctness. But if I am going to give a mark, that is the least part of my marking. The hard part is the part beyond that, the part where the adventure begins.

One other way to rid the curriculum of the poetry nuisance has been considered. More merciful than the others it would neither abolish nor denature the poetry, but only turn it out to disport itself, with the plays and games — in no wise discredited, though given no credit for. Any one who liked to teach poetically could take his subject, whether English, Latin, Greek or French, out into the nowhere along with the poetry. One side of a sharp line would be left to the rigorous and righteous; the other side would be assigned to the flowery where they would know what could be expected of them. Grade marks where more easily given, of course, in the courses concentrating on correctness and exactness as the only forms of honesty recognized by plain people; a general indefinite mark of X in the courses that scatter brains over taste and opinion. On inquiry I have found no teacher willing to take position on either side of the line, either among the rigors or among the flowers. No one is willing to admit that his discipline is not partly in taste and enthusiasm.

How shall a man go through college without having been marked for taste and judgment? What will become of him? What will his end be? He will have to take continuation courses for college graduates. He will have to go to night schools. They are having night schools now, you know, for college graduates. Why? Because they have not been educated enough to find their way around in contemporary literature. They don't know what they may safely like in the libraries and galleries. They don't know how to judge an editorial when they see one. They don't know how to judge a political campaign. They don't know when they are being fooled by a metaphor, an analogy, a parable. And metaphor is, of course, what we are talking about. Education by poetry is education by metaphor.

Suppose we stop short of imagination, initiative, enthusiasm, inspiration and originality — dread words. Suppose we don't mark in such things at all. There are still two minimal things, that we have got to take care of, taste and judgment. Americans are supposed to have more judgment than taste, but taste is there to be dealt with. That is what poetry, the only art in the colleges of arts, is there for. I for my part would not be afraid to go in for enthusiasm. There is the enthusiasm like a blinding light, or the enthusiasm of the deafening shout, the crude enthusiasm that you get uneducated by poetry, outside of poetry. It is exemplified in what I might call 'sunset raving.' You look westward to the sunset, or if you get up early enough, eastward to the sunrise, and you rave. It is oh's and ah's with you and no more.

But the enthusiasm I mean is taken through the prism of the intellect and spread on the screen in a color, all the way from hyperbole at one end — or overstatement, at one end — to understatement at the other end. It is a long strip of dark lines and many colors. Such enthusiasm is one object of all teaching in poetry. I heard wonderful things said about Virgil yesterday, and many of them seemed to me crude enthusiasm, more like a deafening shout, many of them. But one speech had range, something of overstatement, something of statement, and something of understatement. It had all the colors of an enthusiasm passed through an idea.

I would be willing to throw away everything else but that: enthusiasm tamed by metaphor. Let me rest the case there. Enthusiasm tamed to metaphor, tamed to that much of it. I do not think anybody ever knows the discreet use of metaphor, his own and other people's, the discreet handling of metaphor, unless he has been properly educated in poetry.

Poetry begins in trivial metaphors, petty metaphors, 'grace' metaphors, and goes on to the profoundest thinking that we have. Poetry provides the one permissible way of saying one thing and meaning another. People say, 'Why don't you say what you mean?' We never do that, do we, being all of us too much poets. We like to talk in parables and in hints and in indirections — whether from diffidence or some other instinct.

I have wanted in late years to go further and further in making metaphor the whole of thinking. I find some one now and then to agree with me that all thinking, except mathematical thinking, is metaphorical, or all thinking except scientific thinking. The mathematical might be difficult for me to bring in, but the scientific is easy enough.

Once on a time all the Greeks were busy telling each other what the All was — or was like unto. All was three elements, air, earth, and water (we once thought it was ninety elements; now we think it is only one). All was substance, said another. All was change, said a third. But the best and most fruitful was Pythagoras' comparison of the universe with number. Number of what? Number of feet, pounds, and seconds was the answer, and we had science and all that has followed in science. The metaphor has held and held, breaking down only when it came to the spiritual and psychological or the out of the way places of the physical.

The other day we had a visitor here, a noted scientist, whose latest word to the world has been that the more accurately you know where a thing is, the less accurately you are able to state how fast it is moving. You can see why that would be so, without going back to Zeno's problem of the arrow's flight.[1] In carrying numbers into the realm of space and at the same time into the realm of time you are mixing metaphors, that is all, and you are in trouble. They won't mix. The two don't go together.

Let's take two or three more of the metaphors now in use to live by. I have just spoken of one of the new ones, a charming mixed metaphor right in the realm of higher mathematics and higher physics: that the more accurately you state where a thing is, the less accurately you will be able to tell how fast it is moving. And, of course everything is moving. Everything is an event now. Another metaphor. A thing, they say, is an event. Do you believe it is? Not quite. I believe it is almost an event. But I like the comparison of a thing with an event.

I notice another from the same quarter. 'In the neighbourhood of matter space is something like curved.' Isn't that a good one! It seems to me that that is simply and utterly charming — to say that space is something like curved in the neighbourhood of matter. 'Something like.'

Another amusing one is from — what is the book? — I can't say it now; but here is the metaphor. Its aim is to restore you to your ideas of free will. It wants to give you back your freedom of will. All right, here it is on a platter. You know that you can't tell by name what persons in a certain class will be dead ten years after graduation, but you can tell actuarially how many will be dead. Now, just so this scientist says of the particles of matter flying at a screen, striking a screen; you can't tell what individual particles will come, but you can say in general that a certain number will strike in a given time. It shows, you see, that the individual particle can come freely. I asked Bohr[2] about that particularly, and he said, 'Yes, it is so. It can come when it wills and as it wills; and the action of the individual particle is unpredictable. But it is not so of the action of the mass. There you can predict.' He says, 'That gives the individual atom its freedom, but the mass its necessity.'

Another metaphor that has interested us in our time and has done all our thinking for us is the metaphor of evolution. Never mind going into the Latin word. The metaphor is simply the metaphor of the growing plant or of the growing thing. And somebody very brilliantly, quite a while ago, said that the whole universe, the whole of everything, was like unto a growing thing. That is all. I know the metaphor will break down at some point, but it has not failed everywhere. It is a very brilliant metaphor, I acknowledge, though I myself get too tired of the kind of essay that talks about the evolution of candy, we will say, or the evolution of elevators — the evolution of this, that, and the other. Everything is evolution. I emancipate myself by simply saying that I didn't get up the metaphor and so am not much interested in it.

What I am pointing out is that unless you are at home in the metaphor, unless you have had your proper poetical education in the metaphor, you are not safe anywhere. Because you are not at ease with figurative values: you don't know the metaphor in its strength and its weakness. You don't know how far you may expect to ride it and when it may break down with you. You are not safe in science; you are not safe in history. In history, for instance —

to show that is the same in history as elsewhere — I heard somebody say yesterday that Aeneas was to be likened unto (those words, 'likened unto'!) George Washington. He was that type of national hero, the middle-class man, not thinking of being a hero at all, bent on building the future, bent on his children, his descendants. A good metaphor, as far as it goes, and you must know how far. And then he added that Odysseus should be likened unto Theodore Roosevelt. I don't think that is so good. Someone visiting Gibbon at the point of death, said he was the same Gibbon as of old; still at his parallels.

Take the way we have been led into our present position morally, the world over. It is by a sort of metaphorical gradient. There is a kind of thinking — to speak metaphorically — there is a kind of thinking you might say was endemic in the brothel. It is always there. And every now and then in some mysterious way it becomes epidemic in the world. And how does it do so? By using all the good words that virtue has invented to maintain virtue. It uses honesty, first — frankness, sincerity — those words; picks them up, uses them. 'In the name of honesty, let us see what we are.' You know. And then it picks up the word joy. 'Let us in the name of joy, which is the enemy of our ancestors, the Puritans . . . Let us in the name of joy, which is the enemy of the kill-joy Puritan . . .' You see. 'Let us,' and so on. And then, 'In the name of health . . .' Health is another good word. And that is the metaphor Freudianism trades on, mental health. And the first thing we know, it has us all in up to the top knot. I suppose we may blame the artists a good deal, because they are great people to spread by metaphor. The stage too — the stage is always a good intermediary between the two worlds, the under and the upper, if I may say so without personal prejudice to the stage.

In all this, I have only been saying that the devil can quote Scripture, which simply means that the good words you have lying around the devil can use for his purposes as well as anybody else. Never mind about my morality. I am not here to urge anything. I don't care whether the world is good or bad — not on any particular day.

Let me ask you to watch a metaphor breaking down here before you.

Somebody said to me a little while ago, 'It is easy enough for me to think of the universe as a machine, as a mechanism.'

I said, 'You mean the universe is like a machine?'

He said, 'No. I think it is one . . . Well, it is like . . .'

'I think you mean the universe is like a machine.'

'All right. Let it go at that.'

I asked him, 'Did you ever see a machine without a pedal for the foot, or a lever for the hand, or a button for the finger?'

He said 'No — no.'

I said, 'All right. Is the universe like that?'

And he said, 'No. I mean it is like a machine, only . . .'

'. . . it is different from a machine,' I said.

He wanted to go just that far with that metaphor and no further. And so do we all. All metaphor breaks down somewhere. That is the beauty of it. It is touch and go with the metaphor, and until you have lived with it long enough you don't know when it is going. You don't know how much you can get out of it and when it will cease to yield. It is a very living thing. It is as life itself.

I have heard this ever since I can remember, and ever since I have taught: the teacher must teach the pupil to think. I saw a teacher once going around in a great school and snapping pupils' heads with thumb and finger saying, 'Think.' That was when thinking was becoming the fashion. The fashion hasn't yet quite gone out.

We still ask boys in college to think, as in the nineties, but we seldom tell them what thinking means; we seldom tell them it is just putting this and that together; it is saying one thing in terms of another. To tell them is to set their feet on the first rung of a ladder the top of which sticks through the sky.

Greatest of all attempts to say one thing in terms of another is the philosophical attempt to say matter in terms of spirit, or spirit in terms of matter, to make the final unity. That is the greatest attempt that ever failed. We stop just short there. But it is the height of poetry, the height of all thinking, the height of all poetic thinking, that attempt to say matter in terms of spirit and spirit in terms of matter. It is wrong to call anybody a materialist simply because he tries to say spirit in terms of matter, as if that were a sin. Materialism is not the attempt to say all in terms of matter. The only materialist — be he poet, teacher, scientist, politician, or statesman — is the man who gets lost in his material without a gathering metaphor to throw it into shape and order. He is the lost soul.

We ask people to think, and we don't show them what thinking is. Somebody says we don't need to show them how to think; bye and bye they will think. We will give them the forms of sentences and, if they have any ideas, then they will know how to write them. But that is preposterous. All there is to writing is having ideas. To learn to write is to learn to have ideas.

The first little metaphor . . . Take some of the trivial ones. I would rather have trivial ones of my own to live by than the big ones of other people.

I remember a boy saying, 'He is the kind of person that wounds with his shield.' That may be a slender one, of course. It goes a good way in character description. It has poetic grace. 'He is the kind that wounds with his shield.'

The shield reminds me — just to linger a minute — the shield reminds me of the inverted shield spoken of in one of the books of the *Odyssey*, the book that tells about the longest swim on record. I forget how long it lasted — several days, was it? — but at last as Odysseus came near the coast of Phoenicia, he saw it on the horizon 'like an inverted shield.'

There is a better metaphor in the same book. In the end Odysseus comes ashore and crawls up the beach to spend the night under a double olive tree, and it says, as in a lonely farmhouse where it is hard to get fire — I am not quoting exactly — where it is hard to start the fire again if it goes out, they cover the seeds of fire with ashes to preserve it for the night, so Odysseus covered himself with the leaves around him and went to sleep. There you have something that gives you character, something of Odysseus himself. 'Seeds of fire.' So Odysseus covered the seeds of fire in himself. You get the greatness of his nature.

But these are slighter metaphors than the ones we live by. They have their charm, their passing charm. They are as it were the first steps toward the great thoughts, grave thoughts, thoughts lasting to the end.

The metaphor whose manage we are best taught in poetry — that is all there is of thinking. It may not seem far for the mind to go but it is the mind's furthest. The richest accumulation of the ages is the noble metaphors we have rolled up.

I want to add one thing more that the experience of poetry is to anyone who comes close to poetry. There are two ways of coming close to poetry. One is by writing poetry. And some people think I want people to write poetry, but I don't; that is, I don't necessarily. I only want people to write poetry if they want to write poetry. I have never encouraged anybody to write poetry that did not want to write it, and I have not always encouraged those who did want to write it. That ought to be one's own funeral. It is a hard, hard life, as they say.

(I have just been to a city in the West, a city full of poets, a city they have made safe for poets. The whole city is so lovely that you do not have to write it up to make it poetry; it is ready-made for you. But, I don't know — the poetry written in that city might not seem like poetry if read outside of the city. It would be like the jokes made when you were drunk; you have to get drunk again to appreciate them.)

But as I say, there is another way to come close to poetry, fortunately, and that is in the reading of it, not as linguistics, not as history, not as anything but poetry. It is one of the hard things for a teacher to know how close a man has come in reading poetry. How do I know whether a man has come close to Keats in reading Keats? It is hard for me to know. I have lived with some boys a whole year over some of the poets and I have not felt sure whether they have come near what it was all about. One remark sometimes told me. One remark was their mark for the year; had to be — it was all I got that told me what I wanted to know. And that is enough, if it was the right remark, if it came close enough. I think a man might make twenty fool remarks if he made one good one some time in the year. His mark would depend on that good remark.

The closeness — everything depends on the closeness with which you come, and you ought to be marked for the closeness, for nothing else. And that will have to be estimated by chance remarks, not by question and answer. It is only by accident that you know some day how near a person has come.

The person who gets close enough to poetry, he is going to know more about the word *belief* than anybody else knows, even in religion nowadays. There are two or three places where we know belief outside of religion. One of them is at the age of fifteen to twenty, in our self-belief. A young man knows more about himself than he is able to prove to anybody. He has no knowledge that anybody else will accept as knowledge. In his foreknowledge he has something that is going to believe itself into fulfilment, into acceptance.

There is another belief like that, the belief in someone else, a relationship of two that is going to be believed into fulfilment. That is what we are talking about in our novels, the belief of love. And disillusionment that the novels are full of is simply the disillusionment from disappointment in that belief. That belief can fail, of course.

Then there is a literary belief. Every time a poem is written, every time a short story is written, it is written not by cunning, but by belief. The beauty, the something, the little charm of the thing to be, is more felt than known. There is a common jest, one that always annoys me, on the writers, that they write the last end first, and then work up to it; that they lay a train toward one sentence that they think is pretty nice and have all fixed up to set like a trap to close with. No, it should not be that way at all. No one who has ever come close to the arts has failed to see the difference between things written that way, with cunning and device, and the kind that are believed into existence, that begin in something more felt than known. This you can realize quite as well — not quite as well, perhaps, but nearly as well — in reading as you can in writing. I would undertake to separate short stories on that principle; stories that have been believed into existence and stories that have been cunningly devised. And I could separate the poems still more easily.

Now I think — I happen to think — that those three beliefs that I speak of, the self-belief, the love-belief, and the art-belief, are all closely related to the God-belief, that the belief in God is a relationship you can enter into with Him to bring about the future.

There is a national belief like that, too. One feels it. I have been where I came near getting up and walking out on the people who thought that they had to talk against nations, against nationalism, in order to curry favor with internationalism. Their metaphors are all mixed up. They think that because a Frenchman and an American and an Englishman can all sit down on the same platform and receive honors together, it must be that there is no such thing as nations. That kind of bad thinking springs from a source we all know. I should want to say to anyone like that: 'Look! First I want to be a person. And I want you to be a person, and then we can be as interpersonal as you please. We can pull each other's noses — do all sorts of thing. But, first of all, you have got to have the personality. First of all, you have got to have the nations and then they can be as international as they please with each other.'

I should like to use another metaphor on them. I want my palette, if I am a painter, I want my palette on my thumb or on my chair, all clean, pure, separate colors. Then I will do the mixing on the canvas. The canvas is where the work of art is, where we make the conquest. But we want the nations all separate, pure, distinct, things as separate as we can make them; and then in our thoughts, in our arts, and so on, we can do what we please about it.

But I go back. There are four beliefs that I know more about from having lived with poetry. One is the personal belief, which is a knowledge that you don't want to tell other people about because you cannot prove that you know. You are saying nothing about it till you see. The love belief, just the same, has that same shyness. It knows it cannot tell; only the outcome can tell. And the national belief we enter into socially with each other, all together, party of the first part, party of the second part, we enter into that to bring the future of the country. We cannot tell some people what it is we believe, partly, because they are too stupid to understand and partly because we are too proudly vague to explain. And anyway it has got to be fulfilled, and we are not talking until we know more, until we have something to show. And then the literary one in every work of art, not of cunning and craft, mind you, but of real art; that believing the thing into existence, saying as you go more than you even hoped you were going to be able to say, and coming with surprise to an end that you foreknew only with some sort of emotion. And then finally the relationship we enter into with God to believe the future in — to believe the hereafter in.

Poetry [1]

I, too, dislike it: there are things that are important beyond all
                                                      this fiddle.
    Reading it, however, with a perfect contempt for it, one
                                                      discovers in
        it after all, a place for the genuine.
            Hands that can grasp, eyes
            that can dilate, hair that can rise
                if it must, these things are important not because a

high-sounding interpretation can be put upon them but because
                                                      they are
    useful. When they become so derivative as to become
                                                      unintelligible,
        the same thing may be said for all of us, that we
            do not admire what
            we cannot understand: the bat
                holding on upside down or in quest of something to

eat, elephants pushing, a wild horse taking a roll, a tireless wolf
                                                      under
    a tree, the immovable critic twitching his skin like a horse that
                                                      feels a flea, the base-
        ball fan, the statistician —
            nor is it valid
                to discriminate against 'business documents and

school-books'; all these phenomena are important. One must
                                                      make a distinction
    however: when dragged into prominence by half poets, the
                                                      result is not poetry,
        nor till the poets among us can be
            'literalists of
            the imagination' — above
                insolence and triviality and can present

for inspection, 'imaginary gardens with real toads in them,'

shall we have
    it. In the meantime, if you demand on the one hand,
  the raw material of poetry in
    all its rawness and
    that which is on the other hand
      genuine, you are interested in poetry.

Marianne Moore (1887–1972)

VIRGINIA WOOLF

Virginia Woolf (1882–1941), in spasmodic diary entries from the end of 1928 to the end of 1931, chronicles in entrancing detail the completion of one book, the gestation and birth (through all stages of labour) of another, and the conception of a third. *A Writer's Diary*, selected and edited by her husband Leonard Woolf, was published in 1953.

from *A Writer's Diary*

Wednesday, November 28th [1928]

So the days pass and I ask myself sometimes whether one is not hypnotised, as a child by a silver globe, by life; and whether this is living. It's very quick, bright, exciting. But superficial perhaps. I should like to take the globe in my hands and feel it quietly, round, smooth, heavy, and so hold it, day after day. I will read Proust I think. I will go backwards and forwards.

As for my next book, I am going to hold myself from writing till I have it impending in me: grown heavy in my mind like a ripe pear; pendant, gravid, asking to be cut or it will fall. *The Moths* still haunts me, coming, as they always do, unbidden, between tea and dinner, while L.[1] plays the gramophone. I shape a page or two; and make myself stop. Indeed I am up against some difficulties. Fame to begin with. *Orlando* has done very well. Now I could go on writing like that — the tug and suck are at me to do it. People say this was so spontaneous, so natural. And I would like to keep those qualities if I could without losing the others. But those qualities were largely the result of ignoring the others. They came of writing exteriorly; and if I dig, must I not lose them? And what is my own position towards the inner and the outer? I think a kind of ease and dash are good; — yes: I think even externality is good; some combination of them ought to be possible. The idea has come to me that what I want now is to saturate every atom. I mean to eliminate all waste, deadness, superfluity: to give the moment whole; whatever it includes. Say that the moment is a combination of thought; sensation; the voice of the sea. Waste, deadness, come from the inclusion of things that don't belong to the moment; this appalling narrative business of the realist: getting on from lunch to dinner: it is false, unreal, merely conventional. Why admit anything to literature that is not poetry — by which I mean saturated? Is that not my grudge against novelists? that they select nothing? The poets succeeding by simplifying: practically everything is left out. I want to put everything in: yet to saturate. That is what I want to do in *The Moths*. It must include nonsense, fact, sordidity: but made transparent. I think I must read Ibsen and Shakespeare and Racine. And I will write something about them; for that is the best spur, my mind being what it is; then I read with fury and exactness; otherwise I skip and skip; I am a lazy reader. But no: I am surprised and a little disquieted by the remorseless severity of my mind: that it never stops reading and writing; makes me write on Geraldine Jewsbury,[2] on Hardy, on Women — is too professional, too little any longer a dreamy amateur.

Tuesday, December 18th

L. has just been in to consult about a 3rd edition of *Orlando*. This has been ordered; we have sold over 6,000 copies; and sales are still amazingly brisk — 150 today for instance; most days between 50 and 60; always to my surprise. Will they stop or go on? Anyhow my room is secure. For the first time since I married, 1912–1928 — 16 years, I have been spending money. The spending muscle does not work naturally yet. I feel guilty; put off buying, when I know that I should buy; and yet have an agreeable luxurious sense of coins in my pocket beyond my weekly 13/- which was always running out, or being encroached upon.

Thursday, March 28th [1929]

It is a disgrace indeed; no diary has been left so late in the year. The truth was that we went to Berlin on 16th January, and then I was in bed for three weeks afterwards and then could not write, perhaps for another three, and have spent my energy since in one of my excited outbursts of composition — writing what I made up in bed, a final version of *Women and Fiction*.

And as usual I am bored by narrative. . . . One ought to invent a fine narrative style. Certainly there are many new ideas always forming in my head. For one, that I am going to enter a nunnery these next months; and let myself down into my mind; Bloomsbury being done with. I am going to face certain things. It is going to be a time of adventure and attack, rather lonely and painful I think. But solitude will be good for a new book. Of course, I shall make friends. I shall be external outwardly. I shall buy some good clothes and go out into new houses. All the time I shall attack this angular shape in my mind. I think *The Moths* (if that is what I shall call it) will be very sharply cornered. I am not satisfied though with the frame. There is this sudden fertility which may be mere fluency. In old days books were so many sentences absolutely struck with an axe out of crystal: and now my mind is so impatient, so quick, in some ways so desperate.

Sunday, May 12th

Here, having just finished what I call the final revision of *Women and Fiction*[3] so that L. can read it after tea, I stop; surfeited. And the pump, which I was so sanguine as to think ceased, begins again. About *Women and Fiction* I am not sure — a brilliant essay? — I daresay: it has much work in it, many opinions boiled down into a kind of jelly, which I have stained red as far as I can. But I am eager to be off — to write without any boundary coming slick in one's eyes: here my public has been too close; facts; getting them malleable, easily yielding to each other.

Tuesday, May 28th

Now about this book, *The Moths*. How am I to begin it? And what is it to be? I feel no great impulse; no fever; only a great pressure of difficulty. Why write it then? Why write at all? Every morning I write a little sketch, to amuse myself. I am not saying, I might say, that these sketches have any relevance. I am not trying to tell a story. Yet perhaps it might be done in that way. A mind thinking. They might be islands of light — islands in the stream that I am trying to convey; life itself going on. The current of the moths flying strongly this way. A lamp and a flower pot in the centre. The flower can always be changing. But there must be more unity between each scene than I can find at present. Autobiography it might be called. How am I to make one lap, or act, between the coming of the moths, more intense than another; if there are only scenes? One must get the sense that this is the beginning; this is the middle; that the climax — when she opens the window and the moth comes in. I shall have the two different currents — the moths flying along; the flower upright in the centre; a perpetual crumbling and renewing of the plant. In its leaves she might see things happen. But who is she? I am very anxious that she should have no name. I don't want a Lavinia or a Penelope.[4] I want 'she'. But that becomes arty, Liberty greenery yallery [5] somehow: symbolic in loose robes. Of course I can make her think backwards and forwards; I can tell stories. But that's not it. Also I shall do away with exact place and time. Anything may be out of the window — a ship — a desert — London.

Sunday, June 23rd

It was very hot that day, driving to Worthing to see Leonard's mother, my throat hurt me. Next morning I had a headache — so we stayed on at Rodmell till today. At Rodmell I read through *The Common Reader*; and this is very important — I must learn to write more succinctly. Especially in the general idea essays like the last, 'How it strikes a Contemporary,' I am horrified by my own looseness. This is partly that I don't think things out first; partly that I stretch my style to take in crumbs of meaning. But the result is a wobble and diffusity and breathlessness which I detest. One must correct *A Room of One's Own* very carefully before printing. And so I pitched into my great lake of melancholy. Lord how deep it is! What a born melancholic I am! The only way I keep afloat is by working. A note for the summer — I must take more work than I can possibly get done. — No, I don't know what it comes from. Directly I stop working I feel that I am sinking down, down. And as usual I feel that if I sink further I shall reach the truth. That is the only mitigation; a kind of nobility. Solemnity. I shall make myself face the fact that there is nothing — nothing for any of us. Work, reading, writing are all disguises; and relations with people. Yes, even having children would be useless.

However, I now begin to see *The Moths* rather too clearly, or at least strenuously, for my comfort. I think it will begin like this: dawn, the shells on a beach; I don't know — voices of cock and nightingale; and then all the children at a long table — lessons. The beginning. Well, all sorts of characters are to be there. Then the person who is at the table can call out anyone of them at any moment; and build up by that person the mood, tell a story; for instance about dogs or nurses; or some adventure of a child's kind; all to be very Arabian Nights; and so on: this shall be childhood; but it must not be *my* childhood; and boats on the pond; the sense of children; unreality; things oddly proportioned. Then another person or figure must be selected. The unreal world must be round all this — the phantom waves. The Moth must come in; the beautiful single moth. Could one not get the waves to be heard all through? Or the farmyard noises? Some odd irrelevant noises. She might have a book — one book to read in — another to write in — old letters. Early morning light — but this need not be insisted on; because there must be great freedom from 'reality'. Yet everything must have relevance.

Well all this is of course the 'real' life; and nothingness only comes in the absence of this. I have proved this quite certainly in the past half hour. Everything becomes green and vivified in me when I begin to think of *The Moths*. Also, I think, one is much better able to enter into others' ——

Monday, September 10th

Leonard is having a picnic at Charleston[6] and I am here — 'tired'. But why am I tired? Well I am never alone. This is the beginning of my complaint. I am not physically tired so much as psychologically. I have strained and wrung at journalism and proof correction; and underneath has been forming my Moth book. Yes, but it forms very slowly; and what I want is not to write it, but to think it for two or three weeks say — to get into the same current of thought and let that submerge everything. Writing perhaps a few phrases here at my window in the morning. . . .

Really these premonitions of a book — states of soul in creating — are very queer and little apprehended. . . .

And then I am 47: yes; and my infirmities will of course increase. To begin with my eyes. Last year, I think, I could read without spectacles; would pick up a paper and read it in a tube; gradually I found I needed spectacles in bed; and now I can't read a line (unless held at a very odd angle) without them. My new spectacles are much stronger than the old and when I take them off I am blinded for a moment. What other infirmities? I can hear, I think, perfectly: I think I could walk as well as ever. But then will there not be the change of life? And may that not be a difficult and even dangerous time? Obviously one can get over it by facing it with common sense — that it is a natural process; that one can lie out here and read; that one's faculties will be the same afterwards; that one has nothing to worry about in one sense — I've written some interesting books, can make money, can afford a holiday — Oh no; one has nothing to bother about; and these curious intervals in life — I've had many — are the most fruitful artistically — one becomes fertilised — think of my madness at Hogarth — and all the little illnesses — that before I wrote the *Lighthouse* for instance. Six weeks in bed now would make a masterpiece of *Moths*. But that won't be the name. Moths, I suddenly remember, don't fly by day. And there can't be a lighted candle. Altogether, the shape of the book wants considering — and with time I could do it. Here I broke off.

Wednesday, September 25th

Yesterday morning I made another start on *The Moths*, but that won't be its title; and several problems cry out at once to be solved. Who thinks it? And am I outside the thinker? One wants some device which is not a trick.

Friday, October 11th

And I snatch at the idea of writing here in order not to write *Waves* or *Moths* or whatever it is to be called. One thinks one has learnt to write quickly; and one hasn't. And what is odd, I'm not writing with gusto or pleasure: because of the concentration. I am not reeling it off; but sticking it down. Also, never, in my life, did I attack such a vague yet elaborate design; whenever I make a mark I have to think of its relation to a dozen others. And though I could go on easily enough, I am always stopping to consider the whole effect. In particular is there some radical fault in my scheme? I am not quite satisfied with this method of picking out things in the room and being reminded by them of other things. Yet I can't at the moment divine anything which keeps so close to the original design and admits of movement. Hence, perhaps, these October days are to me a little strained and surrounded with silence. . . .

Wednesday, October 23rd

As it is true — I write only for an hour, then rush back feeling I cannot keep my brain on that spin any more — then typewrite, and am done by 12. I will here sum up my impressions before publishing *A Room of One's Own*. It is a little ominous that Morgan won't review it. It makes me suspect that there is a shrill feminine tone in it which my intimate friends will dislike. I forecast, then, that I shall get no criticism, except of the evasive jocular kind, from Lytton, Roger and Morgan;[7] that the press will be kind and talk of its charm and sprightliness; also I shall be attacked for a feminist and hinted at for a Sapphist;[8] Sybil will ask me to luncheon; I shall get a good many letters from young women. I am afraid it will not be taken seriously. Mrs. Woolf is so accomplished a writer that all she says makes easy reading . . . this very feminine logic . . . a book to be put in the hands of girls. I doubt that I mind very much. The Moths; but I think it is to be waves, is trudging along; and I have that to refer to, if I am damped by the other. It is a trifle, I shall say; so it is; but I wrote it with ardour and conviction. . . .

He wrote yesterday, 3 Dec. and said he very much liked it.

Saturday, November 2nd

Oh but I have done quite well so far with *Room of One's Own*; and it sells, I think; and I get unexpected letters. But I am more concerned with my *Waves*. I've just typed out my morning's work; and can't feel altogether sure. There is *something* there (as I felt about *Mrs. Dalloway*) but I can't get at it, squarely; nothing like the speed and certainty of the *Lighthouse: Orlando* mere child's play. Is there some falsity of method, somewhere? Something tricky? — so that the interesting things aren't firmly based? I am in an odd state; feel a cleavage; here's my interesting thing; and there's no quite solid table on which to put it. It might come in a flash, on re-reading — some solvent. I am convinced that I am right to seek for a station whence I can set my people against time and the sea — but Lord, the difficulty of digging oneself in there, with conviction. Yesterday I had conviction; it has gone today.

Saturday, November 30th

I fill in this page, nefariously; at the end of a morning's work. I have begun the second part of *Waves* — I don't know. I don't know. I feel that I am only accumulating notes for a book — whether I shall ever face the labour of writing it, God knows. From some higher station I may be able to pull it together — at Rodmell, in my new room. Reading the *Lighthouse* does not make it easier to write. . . .

Sunday, December 8th

I read and read and finished I daresay 3 foot thick of MS. read carefully too; much of it on the border, and so needing thought. Now, with this load despatched, I am free to begin reading Elizabethans — the little unknown writers, whom I, so ignorant am I, have never heard of, Puttenham, Webb, Harvey. This thought fills me with joy — no overstatement. To begin reading with a pen in my hand, discovering, pouncing, thinking of phrases, when the ground is new, remains one of my great excitements. Oh but L. will sort apples and the little noise upsets me; I can't think what I was going to say. . . .

RODMELL — *Boxing Day*

I find it almost incredibly soothing — a fortnight alone — almost impossible to let oneself have it. Relentlessly we have crushed visitors. We will be alone this once, we say; and really, it seems possible. Then Annie is to me very sympathetic. My bread bakes well. All is rather rapt, simple, quick, effective — except for my blundering on at *The Waves*. I write two pages of arrant nonsense, after straining; I write variations of every sentence; compromises; bad shots; possibilities; till my writing book is like a lunatic's dream. Then I trust to some inspiration on re-reading; and pencil them into some sense. Still I am not satisfied. I think there is something lacking. I sacrifice nothing to seemliness. I press to my centre. I don't care if it all is scratched out. And there is something there. I incline now to try violent shots — at London — at talk — shouldering my way ruthlessly — and then, if nothing comes of it — anyhow I have examined the possibilities. But I wish I enjoyed it more. I don't have it in my head all day like the *Lighthouse* and *Orlando*.

Sunday, January 12th [1930]

Sunday it is. And I have just exclaimed: 'And now I can think of nothing else.' Thanks to my pertinacity and industry, I can now hardly stop making up *The Waves*. The sense of this came acutely about a week ago on beginning to write the *Phantom Party*: now I feel that I can rush on, after 6 months' hacking, and finish: but without the least certainty how it's to achieve any form. Much will have to be discarded: what is essential is to write fast and not break the mood — no holiday, no interval if possible, till it is done. Then rest. Then re-write.

Sunday, January 26th

I am 48: we have been at Rodmell — a wet, windy day again; but on my birthday we walked among the downs, like the folded wings of grey birds; and saw first one fox, very long with his brush stretched; then a second; which had been barking, for the sun was hot over us; it leapt lightly over a fence and entered the furze — a very rare sight. How many foxes are there in England? At night I read Lord Chaplin's life. I cannot yet write naturally in my new room, because the table is not the right height and I must stoop to warm my hands. Everything must be absolutely what I am used to.

I forgot to say that when we made up our 6 months accounts, we found I had made about £3, 020 last year — the salary of a civil servant: a surprise to me, who was content with £200 for so many years. But I shall drop very heavily I think. The Waves won't sell more than 2,000 copies. I am stuck fast in that book — I mean, glued to it, like a fly on gummed paper. Sometimes I am out of touch; but go on; then again feel that I have at last, by violent measures — like breaking through gorse — set my hands on something central.

It has now sold abut 6,500 today, Oct. 30th, 1931 — after 3 weeks. But will stop now, I suppose.

Perhaps I can now say something quite straight out; and at length; and need not be always casting a line to make my book the right shape. But how to pull it together, how to comport it — press it into one — I do not know; nor can I guess the end — it might be a gigantic conversation. The interludes are very difficult, yet I think essential; so as to bridge and also to give a background — the sea; insensitive nature — I don't know. But I think, when I feel this sudden directness, that it must be right: anyhow no other form of fiction suggests itself except as a repetition at the moment.

Sunday, February 16th

To lie on the sofa for a week. I am sitting up today in the usual state of unequal animation. Below normal, with spasmodic desire to write, then to doze. It is a fine cold day and if my energy and sense of duty persist, I shall drive up to Hampstead. But I doubt that I can write to any purpose. A cloud swims in my head. One is too conscious of the body and jolted out of the rut of life to get back to fiction. Once or twice I have felt that odd whirr of wings in the head, which comes when I am ill so often — last year for example at this time I lay in bed constructing *A Room of One's Own* (which sold 10,000 two days ago). If I could stay in bed another fortnight (but there is no chance of that) I believe I should see the whole of *The Waves*. Or of course I might go off on something different. As it is I half incline to insist upon a dash to Cassis; but perhaps this needs more determination than I possess; and we shall dwindle on here. Pinker[9] is walking about the room looking for the bright patch — a sign of spring. I believe these illnesses are in my case — how shall I express it? — partly mystical. Something happens in my mind. It refuses to go on registering impressions. It shuts itself up. It becomes chrysalis. I lie quite torpid, often with acute physical pain — as last year; only discomfort this. Then suddenly something springs. Two nights ago Vita was here; and when she went I began to feel the quality of the evening — how it was spring coming: a silver light; mixing with the early lamps; the cabs all rushing through the streets; I had a tremendous sense of life beginning; mixed with that emotion which is the essence of my feeling, but escapes description (I keep on making up the Hampton Court scene in *The Waves* — Lord how I wonder if I shall pull this book off! It is a litter of fragments so far). Well, as I was saying, between these long pauses, for I am swimming in the head and write rather to stabilise myself than to make a correct statement — I felt the spring beginning; and Vita's life so full and flush; and all the doors opening; and this is I believe the moth shaking its wings in me. I then begin to make up my story whatever it is; ideas rush in me; often though this is before I can control my mind or pen. It is no use trying to write at this stage. And I doubt if I can fill this white monster. I would like to lie down and sleep, but feel ashamed. . . .

Monday, February 17th

And this temperature is up: but it has now gone down; and now

Thursday, February 20th

I must canter my wits if I can. Perhaps some character sketches.

Monday, March 17th

The test of a book (to a writer) is if it makes a space in which, quite naturally, you can say what you want to say. As this morning I could say what Rhoda said. This proves that the book itself is alive: because it has not crushed the thing I wanted to say, but allowed me to slip it in, without any compression or alteration.

Friday, March 28th

Yes, but this book is a very queer business. I had a day of intoxication when I said 'Children are nothing to this': when I sat surveying the whole book complete and quarrelled with L. (about Ethel Smyth) and walked it off, felt the pressure of the form — the splendour, the greatness — as, perhaps I have never felt them. But I shan't race it off in intoxication. I keep pegging away;

and find it the most complex and difficult of all my books. How to end, save by a tremendous discussion, in which every life shall have its voice — a mosaic —— I do not know. The difficulty is that it is all at high pressure. I have not yet mastered the speaking voice. Yet I think something is there; and I propose to go on pegging it down, arduously, and then re-write, reading much of it aloud, like poetry. It will bear expansion. It is compressed I think. It is — whatever I make of it — a large and potential theme — which *Orlando* was not perhaps. At any rate, I have taken my fence.

Wednesday, April 9th

What I now think (about *The Waves*) is that I can give in a very few strokes the essentials of a person's character. It should be done boldly, almost as caricature. I have yesterday entered what may be the last lap. Like every piece of the book it goes by fits and starts. I never get away with it; but am tugged back. I hope this makes for solidity; and must look to my sentences. The abandonment of *Orlando* and *Lighthouse* is much checked by the extreme difficulty of the form — as it was in *Jacob's Room*. I think this is the furthest development so far; but of course it may miss fire somewhere. I think I have kept stoically to the original conception. What I fear is that the re-writing will have to be so drastic that I may entirely muddle it somehow. It is bound to be very imperfect. But I think it possible that I have got my statues against the sky.

Sunday, April 13th

I read Shakespeare *directly* I have finished writing. When my mind is agape and red-hot. Then it is astonishing. I never yet knew how amazing his stretch and speed and word coining power is, until I felt it utterly outpace and outrace my own, seeming to start equal and then I see him draw ahead and do things I could not in my wildest tumult and utmost press of mind imagine. Even the less known plays are written at a speed that is quicker than anybody else's quickest; and the words drop so fast one can't pick them up. Look at this. 'Upon a gather'd lily almost wither'd.' (That is a pure accident. I happen to light on it.) Evidently the pliancy of his mind was so complete that he could furbish out any train of thought; and, relaxing, let fall a shower of such unregarded flowers. Why then should anyone else attempt to write? This is not 'writing' at all. Indeed, I could say that Shakespeare surpasses literature altogether, if I knew what I meant.

Wednesday, April 23rd

This is a very important morning in the history of *The Waves*, because I think I have turned the corner and see the last lap straight ahead. I think I have got

Bernard into the final stride. He will go straight on now, and then stand at the door: and then there will be a last picture of the waves. We are at Rodmell and I daresay I shall stay on a day or two (if I dare) so as not to break the current and finish it. O Lord and then a rest; and then an article; and then back again to this hideous shaping and moulding. There may be some joys in it all the same.

Tuesday, April 29th

And I have just finished, with this very nib-ful of ink, the last sentence of *The Waves*. I should record this for my own information. Yes, it was the greatest stretch of mind I ever knew; certainly the last pages; I don't think they flop as much as usual. And I think I have kept starkly and ascetically to the plan. So much I will say in self-congratulation. But I have never written a book so full of holes and patches; that will need re-building, yes, not only re-modelling. I suspect the structure is wrong. Never mind. I might have done something easy and fluent; and this is a reach after that vision I had, the unhappy summer — or three weeks — at Rodmell, after finishing the *Lighthouse*. (And that reminds me — I must hastily provide my mind with something else, or it will again become pecking and wretched — something imaginative, if possible, and light; for I shall tire of Hazlitt and criticism after the first divine relief; and I feel pleasantly aware of various adumbrations in the back of my head; a life of Duncan; no, something about canvases glowing in a studio; but that can wait.)

P.m.   And I think to myself as I walk down Southampton Row, 'And I have given you a new book.'

Thursday, May 1st

And I have completely ruined my morning. Yes that is literally true. They sent a book from *The Times* as if advised by Heaven of my liberty; and feeling my liberty wild upon me, I rushed to the cable and told Van Doren I would write on Scott. And now having read Scott, or the editor whom Hugh [10] provides, I won't and can't ; and have got into a fret trying to read it, and writing to Richmond to say I can't: have wasted the brilliant first of May which makes my skylight blue and gold; have only a rubbish heap in my head; can't read and can't write and can't think. The truth is, of course, I want to be back at *The Waves*. Yes that is the truth. Unlike all my other books in every way, it is unlike them in this, that I begin to re-write it, or conceive it again with ardour, directly I have done. I begin to see what I had in my mind; and want to begin cutting out masses of irrelevance and clearing, sharpening and making the good phrases shine. One wave after another. No room. And so on. But then we are going touring Devon and Cornwall on Sunday, which means a week off; and then I shall perhaps make my critical brain do a month's work for exercise. What could it be set to? Or a story? — no, not another story now . . .

Wednesday, August 20th

The Waves is I think resolving itself (I am at page 100) into a series of dramatic soliloquies. The thing is to keep them running homogeneously in and out, in the rhythm of the waves. Can they be read consecutively? I know nothing about that. I think this is the greatest opportunity I have yet been able to give myself; therefore I suppose the most complete failure. Yet I respect myself for writing this book — yes — even though it exhibits my congenital faults. . . .

Friday, December 12th

This, I think, is the last day's breathing space I allow myself before I tackle the last lap of *The Waves*. I have had a week off — that is to say I have written three little sketches and dawdled and spent a morning shopping and a morning, this morning, arranging my new table and doing odds and ends — but I think I have got my breath again and must be off for three or perhaps four weeks more. Then, as I think, I shall make one consecutive writing of *The Waves* etc. — the interludes — so as to work it into one — and then, oh dear, some must be written again; and then, corrections; and then send to Mabel; and then correct the type; and then give to Leonard. Leonard perhaps shall get it some time late in March. Then put away; then print, perhaps in June.

Monday, December 22nd

It occurred to me last night while listening to a Beethoven quartet that I would merge all the interjected passages into Bernard's final speech and end with the words O solitude: thus making him absorb all those scenes and having no further break. This is also to show that the theme effort, effort, dominates: not the waves: and personality: and defiance: but I am not sure of the effect artistically; because the proportions may need the intervention of the waves finally so as to make a conclusion. . . .

Tuesday, December 30th

What it wants is presumably unity; but it is I think rather good (I am talking to myself over the fire about *The Waves*). Suppose I could run all the scenes together more? — by rhythms chiefly. So as to avoid those cuts; so as to make the blood run like a torrent from end to end — I don't want the waste that the breaks give; I want to avoid chapters; that indeed is my achievement, if any, here: a saturated unchopped completeness; changes of scene, of mind, of person, done without spilling a drop. Now if it could be worked over with heat and currency, that's all it wants. And I am getting my blood up (temp. 99). But all the same I went into Lewes and the Keynes came to tea, and having got astride my saddle the whole world falls into shape; it is this writing that gives me my proportions.

Wednesday, January 7th [1931]

My head is not in the first spring of energy: this fortnight has brought me no views of the lapping downs — no fields and hedges — too many firelit houses and lit up pages and pen and ink — curse my influenza. It is very quiet here — not a sound but the hiss of the gas. Oh but the cold was too great at Rodmell. I was frozen like a small sparrow. And I did write a few staggering sentences. Few books have interested me more to write than *The Waves*. Why, even now, at the end, I'm turning up a stone or two: no glibness, no assurance; you see, I could perhaps do B.'s soliloquy in such a way as to break up, dig deep, make prose move — yes I swear — as prose has never moved before; from the chuckle, the babble to the rhapsody. Something new goes into my pot every morning — something that's never been got at before. The high wind can't blow, because I'm chopping and tacking all the time. And I've stored a few ideas for articles: one on Gosse — the critic, as talker: the armchair critic; one on Letters — one on Queens.

Now this is true: *The Waves* is written at such high pressure that I can't take it up and read it through between tea and dinner; I can only write it for about one hour, from 10 to 11.30. And the typing is almost the hardest part of the work. Heaven help me if all my little 80,000 word books are going in future to cost me two years! But I shall fling off, like a cutter leaning on its side, on some swifter, slighter adventure — another *Orlando* perhaps.

Tuesday, January 20th

I have this moment, while having my bath, conceived an entire new book[11] — a sequel to *A Room of One's Own* — about the sexual life of women: to be called Professions for Women perhaps — Lord how exciting! This sprang out of my paper to be read on Wednesday to Pippa's society. Now for *The Waves*. Thank God — but I'm very much excited.

(This is *Here and Now*, I think. May '34.)

Friday, January 23rd

Too much excited, alas, to get on with *The Waves*. One goes on making up 'The Open Door,' or whatever it is to be called. The didactive demonstrative style conflicts with the dramatic: I find it hard to get back inside Bernard again.

Thursday, January 26th [sic]

Heaven be praised, I can truthfully say on this first day of being 49 that I have shaken off the obsession of *Opening the Door*, and have returned to *Waves*: and have this instant seen the entire book whole, and now I can finish it — say in under 3 weeks. . . .

Monday, February 2nd

I think I am about to finish *The Waves*. I think I might finish it on Saturday.

This is merely an author's note: never have I screwed my brain so tight over a book. The proof is that I am incapable of other reading or writing. I can only flop once the morning is over. Oh Lord the relief when this week is over and I have at any rate the feeling that I have wound up and done with that long labour: ended that vision. I think I have just done what I meant; of course I have altered the scheme considerably; but my feeling is that I have insisted upon saying, by hook or by crook, certain things I meant to say. I imagine that the hookedness may be so great that it will be a failure from a reader's point of view. Well, never mind: it is a brave attempt. I think, something struggled for. Oh and then the delight of skirmishing free again — the delight of being idle and not much minding what happens; and then I shall be able to read again, with all my mind — a thing I haven't done these four months I daresay. This will have taken me 18 months to write: and we can't publish it till the autumn I suppose.

Wednesday, February 4th

A day ruined, for us both. L. has to go every morning at 10.15 to the Courts, where his jury is still called, but respited always till 10.15 the next day; and this morning, which should have dealt a formidable blow at *The Waves* — B. is within two days I think of saying O Death — was ruined by Elly,12 who was to have come at 9.30 sharp but did not come till 11. And now it is 12.30 and we sat talking about the period and professional women, after the usual rites with the stethoscope, seeking vainly the cause of my temperature. If we like to spend 7 guineas we might catch a bug — but we don't like. And so I am to eat Bemax and — the usual routine.

How strange and wilful these last exacerbations of *The Waves* are! I was to have finished it at Christmas. . . .

Saturday, February 7th

Here in the few minutes that remain, I must record, heaven be praised, the end of *The Waves*. I wrote the words O Death fifteen minutes ago, having reeled across the last ten pages with some moments of such intensity and intoxication that I seemed only to stumble after my own voice, or almost, after some sort of speaker (as when I was mad) I was almost afraid, remembering the voices that used to fly ahead. Anyhow, it is done; and I have been sitting these 15 minutes in a state of glory, and calm, and some tears, thinking of Thoby[13] and if I could write Julian Thoby Stephen 1881 – 1906 on the first page. I suppose not. How physical the sense of triumph and relief is! Whether good or bad, it's done; and, as I certainly felt at the end, not merely finished, but rounded off, completed, the thing stated — how hastily, how fragmentarily I know; but I mean that I have netted that fin in the waste of water which appeared to me over the marshes out of my window at Rodmell when I was coming to an end of *To the Lighthouse*.

What interests me in the last stage was the freedom and boldness with which my imagination picked up, used and tossed aside all the images, symbols which I had prepared. I am sure that this is the right way of using them — not in set pieces, as I had tried at first, coherently, but simply as images, never making them work out; only suggest. Thus I hope to have kept the sound of the sea and the birds, dawn and garden subconsciously present, doing their work under ground. . . .

Saturday, April 11th

Oh I am so tired of correcting my own writing — these 8 articles — I have however learnt I think to dash: not to finick. I mean the writing is free enough; it's the repulsiveness of correcting that nauseates me. And the cramming in and the cutting out. And articles and more articles are asked for. Forever I could write articles.

But I have no pen — well, it will make just a mark. And not much to say, or rather too much and not the mood.

Wednesday, May 13th

Unless I write a few sentences here from time to time I shall, as they say, forget the use of my pen. I am now engaged in typing out from start to finish the 332 pages of that very condensed book *The Waves*. I do 7 or 8 daily; by which means I hope to have the whole complete by June 16th or thereabouts.

This requires some resolution; but I can see no other way to make all the corrections and keep the lilt and join up and expand and do all the other final processes. It is like sweeping over an entire canvas with a wet brush.

Saturday, May 30th

No, I have just said, it being 12.45, I cannot write any more, and indeed I
p.162. cannot: I am copying the death chapter; have re-written it
therefore twice. I shall go at it again and finish it, I hope, this afternoon.
halfway in But how it rolls into a tight ball the muscles in my brain! This
26 days. is the most concentrated work I have ever done — and oh the
Shall finish relief when it is finished. But also the most interesting.
by 1st July
with luck.

Tuesday, June 23rd

And yesterday, 22nd June, when, I think, the days begin to draw in, I finished my re-typing of *The Waves*. Not that it is finished — oh dear no. For then I must correct the re-re-typing. This work I began on May 5th, and no one can say that I have been hasty or careless this time; though I doubt not the lapses and slovenliness are innumerable.

Tuesday, July 7th

O to seek relief from this incessant correction (I am doing the interludes) and write a few words carelessly. Still better, to write nothing; to tramp over the downs, blown like thistle, as irresponsible. And to get away from this hard knot in which my brain has been so tight spun — I mean *The Waves*. Such are my sentiments at half past twelve on Tuesday July 7th — a fine day I think — and everything, so the tag runs in my head, handsome about us.

Tuesday, July 14th

I had meant to say that I have just finished correcting the Hampton Court scene. (This is the final correction, please God!)
    But my *Waves* account runs, I think, as follows:—
    I began it, seriously, about September 10th 1929.
    I finished the first version on April 10th 1930.
    I began the second version on May 1st 1930.
    I finished the second version on February 7th 1931.
    I began to correct the second version on May 1st 1931, finished 22nd June 1931.
    I began to correct the typescript on 25th June 1931.

Shall finish (I hope) 18th July 1931.
Then remain only the proofs.

Friday, July 17th

Yes, this morning I think I may say I have finished. That is to say I have once
more, for the 18th time, copied out the opening sentences. L.
will read it tomorrow; and I shall open this book to record his
verdict. My own opinion — oh dear — it's a difficult book. I
don't know that I've ever felt so strained. And I'm nervous, I
confess, about L. For one thing he will be honest, more than
usually. And it may be a failure. And I can't do any more. And I'm inclined
to think it good but incoherent, inspissate; one jerk succeeding another.
Anyhow it is laboured, compact. Anyhow I had a shot at my vision — if it's
not a catch, it's a cast in the right direction. But I'm nervous. It may be small
and finicky in general effect. Lord knows. As I say, repeating it to enforce the
rather unpleasant little lift in my heart, I shall be nervous to hear what L. says
when he comes out, say tomorrow night or Sunday morning, to my garden
room, carrying the MS. and sits himself down and begins 'Well!'

Which I then lost.

Sunday, July 19th

'It is a masterpiece,' said L., coming out to my lodge this morning. 'And the
best of your books.' This note I make; adding that he also thinks the first 100
pages extremely difficult and is doubtful how far any common reader will
follow. But Lord! what a relief! I stumped off in the rain to make a little
round to Rat Farm in jubilation and am almost resigned to the fact that a goat
farm, with a house to be built, is now in process on the slope near Northease.

Monday, August 10th

I have now — 10.45 — read the first chapter of *The Waves*, and made no
changes, save 2 words and 3 commas. Yes, anyhow this is exact and to the
point. I like it. And see that for once my proofs will be despatched with a few
pencil strokes. Now my blood mounts: I think 'I am taking my fences . . .
We have asked Raymond. I am forging through the sea, in spite of headache,
in spite of bitterness. I may also get a   .'[14] I will now write a little
at *Flush*.

Saturday, August 15th

I am in rather a flutter — proof reading. I can only read a few pages at a time.
So it was when I wrote it and Heaven knows what virtue it has, this ecstatic
book.

Sunday, August 16th

I should really apologise to this book for using it as I am doing to write off my aimlessness; that is I am doing my proofs — the last chapter this morning — and find that I must stop after half an hour and let my mind spread, after these moments of concentration. I cannot write my life of *Flush*, because the rhythm is wrong. I think *The Waves* is anyhow tense and packed; since it screws my brain up like this. And what will the reviewers say? And my friends? They can't, of course, find anything very new to say.

Monday, August 17th

Well now, it being just after 12.30, I have put the last corrections in *The Waves*, done my proofs; and they shall go tomorrow — never, never to be looked at again by me, I imagine.

Tuesday, September 22nd

And Miss Holtby says 'It is a poem, more completely than any of your books, of course. It is most rarely subtle. It has seen more deeply into the human heart, perhaps, than even *To the Lighthouse* . . .' and though I copy the sentence, because it is the chart of my temperature, Lord, as I say, that temperature which was deathly low this time last week and then fever high, doesn't rise: is normal. I suppose I'm safe; I think people can only repeat. And I've forgotten so much. What I want is to be told that this is solid and means something. What it means I myself shan't know till I write another book. And I'm the hare, a long way ahead of the hounds my critics.

Monday, October 5th

A note to say I am all trembling with pleasure — can't go on with my Letter — because Harold Nicolson has rung up to say *The Waves* is a masterpiece. Ah Hah — so it wasn't all wasted then. I mean this vision I had here has some force upon other minds. Now for a cigarette and then a return to sober composition.

Well, to continue this egotistic diary: I am not terribly excited; no; at arms length more than usual; all this talk, because if the *W.* is anything it is an adventure which I go on alone; and the dear old *Lit. Sup*: who twinkles and beams and patronises — a long, and for *The Times*, kind and outspoken review — don't stir me very much. Nor Harold in *Action* either. Yes; to some extent; I should have been unhappy had they blamed, but Lord, how far away I become from all this; and we're jaded too, with people, with doing up parcels. I wonder if it is good to feel this remoteness — that is, that *The Waves* is not what they say. Odd, that they (*The Times*) should praise my characters when I meant to have none. But I'm jaded; I want my marsh, my down, a quiet waking in my airy bedroom. Broadcasting tonight; to Rodmell tomorrow. Next week I shall have to stand the racket.

Friday, October 9th

Really, this unintelligible book is being better 'received' than any of them. A note in *The Times* proper — the first time this has been allowed me. And it sells — how unexpected, how odd that people can read that difficult grinding stuff!

Saturday, October 17th

More notes on *The Waves*. The sales, these past three days, have fallen to 50 or so: after the great flare up when we sold 500 in one day, the brushwood has died down, as I foretold. (Not that I thought we should sell more than 3,000.) What has happened is that the library readers can't get through it and are sending their copies back. So, I prophesy, it will now dribble along till we have sold 6,000 and then almost die, yet not quite. For it has been received, as I may say, quoting the stock phrases without vanity, with applause. All the provinces read enthusiastically. I am rather, in a sense, as the M.'s would say, touched. The unknown provincial reviewers say with almost one accord, here is Mrs. Woolf doing her best work; it can't be popular; but we respect her for so doing; and find *The Waves* positively exciting. I am in danger, indeed, of becoming our leading novelist, and not with the highbrows only.

Monday, November 16th

Here I will give myself the pleasure — shall I? — of copying a sentence or two from Morgan's unsolicited letter on *The Waves*:—

'I expect I shall write to you again when I have re-read *The Waves*. I have been looking in it and talking about it at Cambridge. It's difficult to express oneself about a work which one feels to be so very important, but I've the sort of excitement over it which comes from believing that one's encountered a classic.'

I daresay that gives me more substantial pleasure than any letter I've had about any book. Yes, I think it does, coming from Morgan. For one thing it gives me reason to think I shall be right to go on along this very lonely path. I mean in the City today I was thinking of another book — about shopkeepers, and publicans, with low life scenes: and I ratified this sketch by Morgan's judgement. Dadie[15] agrees too. Oh yes, between 50 and 60 I think I shall write out some very singular books, if I live. I mean I think I am about to embody at last the exact shapes my brain holds. What a long toil to reach this beginning — if *The Waves* is my first work in my own style! To be noted, as curiosities of my literary history: I sedulously avoid meeting Roger and Lytton whom I suspect do not like *The Waves*.

I am working very hard — in my way, to furbish up two long Elizabethan articles to front a new *Common Reader*: then I must go through the whole long list of those articles. I feel too, at the back of my brain, that I can devise a new critical method; something far less stiff and formal than these *Times* articles. But I must keep to the old style in this volume. And how, I wonder, could I do it? There must be simpler, subtler, closer means of writing about books, as about people, could I hit upon it. (*The Waves* has sold more than 7,000.)

E. P. Ode pour l'Election de son Sépulchre

For three years, out of key with his time,
He strove to resuscitate the dead art
Of poetry; to maintain 'the sublime'
In the old sense. Wrong from the start —

No, hardly, but seeing he had been born
In a half savage country, out of date;
Bent resolutely on wringing lilies from the acorn;
Capaneus;[1] trout for factitious bait;

῎Ιδμεν γάρ τοι πάνθ᾽, ὅσ᾽ ἐνὶ Τροίῃ [2]
Caught in the unstopped ear;
Giving the rocks small lee-way
The chopped seas held him, therefore, that year.

His true Penelope[3] was Flaubert,[4]
He fished by obstinate isles;
Observed the elegance of Circe's[5] hair
Rather than the mottoes on sun-dials.

Unaffected by 'the march of events,'
He passed from men's memory in *l'an trentiesme
De son eage*;[6] the case presents
No adjunct to the Muses' diadem.

Ezra Pound (1885–1972)

EZRA POUND

Ezra Pound (1885–1972) ranks as one of the two or three most original and influential poetic voices in twentieth century English writing. The idiosyncratic economy of this extract illustrates and reinforces the points he is making, while his 'tests and composition exercises' anticipate modern 'peer-editing' techniques for the teaching of writing.

from *ABC of Reading*

Coming round again to the starting-point.

Language is a means of communication. To charge language with meaning to the utmost possible degree, we have, as stated, the three chief means:

I throwing the object (fixed or moving) on to the visual imagination.

II inducing emotional correlations by the sound and rhythm of the speech.

III inducing both of the effects by stimulating the associations (intellectual or emotional) that have remained in the receiver's consciousness in relation to the actual words or word groups employed.

(phanopoeia, melopoeia, logopoeia)[1]

Incompetence will show in the use of too many words.

The reader's first and simplest test of an author will be to look for words that do not function; that contribute nothing to the meaning OR that distract from the MOST important factor of the meaning to factors of minor importance.

.   .   .   .   .

One definition of beauty is: aptness to purpose.

Whether it is a good definition or not, you can readily see that a good deal of BAD criticism has been written by men who assume that an author is trying to do what he is NOT trying to do.

Incredible as it now seems, the bad critics of Keats' time found his writing 'obscure,' which meant that they couldn't understand WHY Keats wrote.

Most human perceptions date from a long time ago, or are derivable from perceptions that gifted men have had long before we were born. The race discovers, and rediscovers.

TESTS AND COMPOSITION EXERCISES

I

1 Let the pupils exchange composition papers and see how many and what useless words have been used — how many words that convey nothing new.

2 How many words that obscure the meaning.

3 How many words out of their usual place, and whether this alteration makes the statement in any way more interesting or energetic.

4 Whether a sentence is ambiguous; whether it really means more than one thing or more than the writer intended; whether it can be so read as to mean something different.

5 Whether there is something clear on paper, but ambiguous if spoken aloud.

II

It is said that Flaubert taught De Maupassant to write. When De Maupassant returned from a walk Flaubert would ask him to describe someone, say a concierge whom they would both pass in their next walk, and to describe the person so that Flaubert would recognize, say, the concierge and not mistake her for some other concierge and not the one De Maupassant had described.

SECOND SET

1 Let the pupil write the description of a tree.

2 Of a tree without mentioning the name of the tree (larch, pine, etc.) so that the reader will not mistake it for the description of some other kind of tree.

3 Try some object in the class-room.

4 Describe the light and shadow on the school-room clock or some other object.

5 If it can be done without breach of the peace, the pupil could write descriptions of some other pupil. The author suggests that the pupil should not describe the instructor, otherwise the description might become a vehicle of emotion, and subject to more complicated rules of composition than the class is yet ready to cope with.

In all these descriptions the test would be accuracy and vividness, the pupil receiving the other's paper would be the gauge. He would recognize or not recognize the object or person described.

Rodolfo Agricola in an edition dating from fifteen hundred and something says one writes: *ut doceat, ut moveat ut delectet*, to teach, to move or to delight.

.   .   .   .   .

A great deal of bad criticism is due to men not seeing which of these three motives underlies a given composition.

The converse processes, not considered by the pious teachers of antiquity, would be to obscure, to bamboozle or mislead, and to bore.

The reader or auditor is at liberty to remain passive and submit to these operations if he so choose.

FURTHER TESTS

Let the pupil examine a given piece of writing, say, the day's editorial in a newspaper, to see whether the writer is trying to conceal something; to see whether he is 'veiling his meaning'; whether he is afraid to say what he thinks; whether he is trying to appear to think without really doing any thinking.

Metrical writing

1 Let the pupil try to write in the metre of any poem he likes.

2 Let him write words to a well-known tune.

3 Let him try to write words to the same tune in such a way that the words will not be distorted when one sings them.

4 Let the pupil write a poem in some strophe[2] form he likes.

5 Let him parody some poem he finds ridiculous, either because of falsity in the statement, or falsity in the disposition of the writer, or for pretentiousness,

of one kind or another, or for any other reason that strikes his risible faculties, his sense of irony.

The gauging pupil should be asked to recognize what author is being parodied. And whether the joke is on the parodied or the parodist. Whether the parody exposes a real defect, or merely makes use of an author's mechanism to expose a more trivial contents.

Note: No harm has ever yet been done a good poem by this process. FitzGerald's *Rubaiyat* has survived hundreds of parodies, that are not really parodies either of Omar or FitzGerald, but only poems written in that form of strophe.

Note: There is a tradition that in Provence it was considered plagiarism to take a man's form, just as it is now considered plagiarism to take his subject matter or plot.

Poems frankly written to another man's strophe form or tune were called 'Sirventes,' and were usually satirical.

FURTHER TESTS

1 Let the pupils in exchanging themes judge whether the theme before them really says anything.

2 Let them judge whether it tells them anything or 'makes them see anything' they hadn't noticed before, especially in regard to some familiar scene or object.

3 Variant: whether the writer really had to KNOW something about the subject or scene before being able to write the page under consideration.

The question of a word or phrase being 'useless' is not merely a numerical problem.

Anatole France in criticizing French dramatists pointed out that on the stage, the words must give time for the action; they must give time for the audience to take count of what is going on.

Even on the printed page there is an analogous case.

Tacitus in writing Latin can use certain forms of condensation that don't necessarily translate advantageously into English.

The reader will often misjudge a condensed writer by trying to read him too fast.

The secret of popular writing is never to put more on a given page than the common reader can lap off it with no strain WHATSOEVER on his habitually slack attention.

Anatole France is said to have spent a great deal of time searching for the *least possible* variant that would turn the most worn-out and commonest phrases of journalism into something distinguished.

Such research is sometimes termed 'classicism.'

This is the greatest possible remove from the usual English stylist's trend or urge toward a style different from everyone else's.

Ars Poetica

A poem should be palpable and mute
As a globed fruit,

Dumb
As old medallions to the thumb,

Silent as the sleeve-worn stone
Of casement ledges where the moss has grown —

A poem should be wordless
As the flight of birds.

*

A poem should be motionless in time
As the moon climbs,

Leaving, as the moon releases
Twig by twig the night-entangled trees,

Leaving, as the moon behind the winter leaves,
Memory by memory the mind —

A poem should be motionless in time
As the moon climbs.

*

A poem should be equal to:
Not true.

For all the history of grief
An empty doorway and a maple leaf.

For love
The leaning grasses and two lights above the sea —

A poem should not mean
But be.

Archibald MacLeish (1892–1982)

KATHERINE MANSFIELD

Katherine Mansfield (1888–1923) left her native New Zealand in 1903 for Europe, where she was to spend most of her life. She habitually filled her *Journal* with comments on people and events rather than on her own writings. The entries from July to November 1921, however, reveal something of the rigorous and meticulous criteria by which she judged her own performance. The sub-headings and square-bracketed explanatory comments are the work of her husband John Middleton Murry, who appears in her *Journal* as J., and after her death edited it for publication.

from *The Journal*

*J*uly. I finished *Mr. and Mrs. Dove* yesterday. I am not altogether pleased with it. It's a little bit made up. It's not inevitable. I mean to imply that those two may not be happy together — that that is the kind of reason for which a young girl marries. But have I done so? I don't think so. Besides, it's not *strong* enough. I want to be nearer — far, far nearer than that. I want to use all my force even when I am taking a fine line. And I have a sneaking notion that I have, at the end, used the Doves *unwarrantably*. *Tu sais ce que je veux dire.*[1] I used them to round off something — didn't I? Is that quite my game? No, it's not. It's not quite the kind of truth I'm after. Now for *Susannah*. All must be *deeply felt*.

But what is one to do, with this wretched cat and mouse act? There's my difficulty! I must try to write this afternoon instead. There is no reason why I shouldn't! No reason, except the after-effects of pain on a weakened organism.

July 23. Finished *An Ideal Family* yesterday. It seems to me better than *The Doves*, but still it's not good enough. I worked at it hard enough, God knows, and yet I didn't get the deepest truth out of the idea, even once. What *is* this feeling? I feel again that this kind of knowledge is too easy for me; it's even a kind of trickery. I know so much more. This looks and smells like a story, but I wouldn't buy it. I don't want to possess it — to live with it. NO. Once I have written two more, I shall tackle something different — a long story: *At the Bay*, with more difficult relationships. That's the whole problem.

"Out of the pocket of the mackintosh she took an ample bag, which she opened and peered into and shook. Her eyebrows were raised, her lips pressed together. . . ."

"And a very long shining blue-black hairpin gleaming on the faded carpet. . . ."

"She shuddered. And now when she looked at his photograph, even the white flower in his buttonhole looked as though it were made of a curl of mutton-fat. . . ."

"And she saw Mr. Bailey in a blue apron standing at the back of one of those horrible shops. He had one hand on his hip, the other grasped the handle of a long knife that was stuck into a huge chopping block. At the back of him there hung a fringe of small rabbits, their feet tied together, a dark clot of blood trembling from their noses. . . ."

July 18. The noise in this house this morning is sheer hell. It has gone on steadily since shortly after six o'clock, and for some reason the maid seems to have completely lost her head. It's now nearly ten, and she hasn't cleared the breakfast away. I have to go again to the Palace at 11, and the consequence is I'm rather nervous anyway. And I've had the flowers to do and various things to see to like — laundry. I can hardly bear it. Now she plods up. Bang! She will be at the door in a moment. I don't know how to stand it if it goes on. She's here. She's about to put the things in the lift. What are her thoughts? I don't know or care. But I bitterly long for a little private room where I can work undisturbed. The balcony is not good enough; neither is this *salon*. Here again, J. has beaten me. And it's not half so important for him. . . .

A Welcome.

And because, when you arrive unexpected, there is so often a cold gleam in the hussif's eye which means: "I can manage the sheets perfectly, but the blankets are certainly going to be a problem," I would have you met in the doorway by a young creature carrying a not too bright lamp, it being, of course, late evening, and chanting, as you brush under the jasmine porch:

> Be not afraid, the house is full of blankets,
> Red ones and white ones, lovely beyond dreaming,
> Key-pattern, tasselled, camel-hair and woolly,
> Softer than sleep or the bosom of a swan.[2]

[In the middle of the manuscript of *Her First Ball*.]

July 25. All this! All that I write — all that I am — is on the border of the sea. It's a kind of playing. I want to put *all* my force behind it, but somehow, I *cannot*! . . .

August. "I have been writing a story about an old man."

She looked vague. "But I don't think I like old men — do you?" said she. "They *exude* so."

This horrified me. It seemed so infernally petty, and more than that . . . it was the saying of a vulgar little mind.

Later. I think it was shyness.

August 11. I don't know how I may write this next story. It's so difficult. But I suppose I shall. The trouble is I am so infernally cold.

[The "next story" was *The Voyage.* The finished manuscript is dated August 14, 1921.]

[From an unposted letter.]

I would have written a card before, but I have been — am — ill, and to-day's the first day I've taken a pen even so far. I've had an attack of what the doctor calls acute enteritis. I think it was poisoning. Very high fever and sickness and dysentery and so on. *Horrible.* I decided yesterday to go to the Palace, but to-day makes me feel I'll try to see it out here. J. is awfully kind in the menial offices of nurse, and I've not been able to take any food except warm milk, so Ernestine can't work her worst on me. She seems, poor creature, to be much more stupid than ever! Burns everything! Leaves us without eggs, and went off for her afternoon yesterday without a word. We didn't even know she was gone.

Love.

August. A sudden idea of the relationship between 'lovers.'

We are neither male nor female. We are a compound of both. I choose the male who will develop and expand the male in me; he chooses me to expand the female in him. Being made 'whole.' Yes, but that's a process. By love serve ye one another... And why I choose *one* man for this rather than many is for safety. We bind ourselves within a ring and that ring is as it were a wall against the outside world. It is our refuge, our shelter. Here the tricks of life will not be played. Here is *safety* for us to *grow.*

Why, I talk like a child!

August 29. "If I could only sweep all my garden up the hill, to your doors!" Her perfect little gesture as she said this.

The Candlestick.

[An imaginary letter.]

Many thanks for your stuffy letter. As for the candlestick, dear, if you remember, I gave it you on your last birthday. No wonder it reminded you of me. I have kept it in its paper and intend to return it to you with a pretty little note on your next. Or shall I first send it to you as an early Christmas present and do you return it as a late one or a New Year's gift. Easter we shall leave out. It would be a trifle excessive at Easter. I wonder which of us will be in possession of it at the last. If it is on my side, I shall leave it to you in my will, all proper, and I think it would be nice of you, Camilla, to desire that it should be buried with you. Besides, one's mind faints at the idea of a candlestick whirling through space and time for ever — a *fliegende*[3] candlestick, in fact!

I have been suffering from wind round the heart. Such a tiresome complaint, but not dangerous. Really, for anything to be so painful I think I would prefer a spice of danger added. The first act was brought by a fit of laughing.

September. September is different from all other months. It is more magical. I feel the strange chemical change in the earth which produces mushrooms is the cause, too, of this extra 'life' in the air — a resilience, a sparkle. For days the weather has been the same. One wakes to see the trees outside bathed in green-gold light. It's fresh — not cold. It's clear. The sky is a light pure blue. During the morning the sun gets hot. There is a haze over the mountains. Occasionally a squirrel appears, runs up the mast of a pine-tree, seizes a cone and sits in the crook of a branch, holding it like a banana. Now and again a little bird, hanging upside-down, pecks at the seed. There is a constant sound of bells from the valley. It keeps on all day, from early to late.

Midday — with long shadows. Hot and still. And yet there's always that taste of a berry rather than scent of a flower in the air. But what can one say of the afternoons? Of the evening? The rose, the gold on the mountains, the quick mounting shadows? But it's soon cold — Beautifully cold, however.

September.

[The following occurs in the middle of an unpublished and unfinished MS. called "By Moonlight." "Karori" was the "novel" of which *Prelude* and *At the Bay* were — at one time — to have formed parts. But eventually the idea was abandoned, because K.M. saw that her "novel" would have been so unlike a novel that it was no use calling it one.]

I am stuck beyond words, and again it seems to me that what I am doing has *no form*! I ought to finish my book of *stories first* and then, when it's gone, really get down to my novel, *Karori*.

Why should I be so passionately determined to disguise this, I don't quite know. But here I lie, pretending, as Heaven knows how often I have before, to write. Supposing I were to give up this pretence and really did try? Supposing I only wrote half a page in a day — it would be half a page to the good; and I should at least be training my mind to get into the habit of regular performance. As it is, every day sees me further off my goal. *And,* once I had this book finished, I'm free to start the real one. *And* it's a question of money.

But my idea, even of the short story, has changed rather, lately — That was lucky! J. opened the door softly and I was apparently really truly engaged. . . . And — no, enough of this. It has served its purpose. It has put me on the right lines.

[At the end of the same MS. is this note]

This isn't bad, but at the same time it's not good. It's too easy. . . . I wish I could go back to N.Z. for a year. But I can't possibly just now. I don't see why not, in two years' time though.

[An unposted letter.]

October 13. Dear *Friend.* I like your criticism. It is right you should have hated those things in me. For I was careless and false. I was not *true* in those days. But I have been trying for a long time now to "squeeze the slave out of my soul." . . . I just want to let you know.

Oh, I am in the middle of a nice story [*The Garden Party*]. I wish you would like it. I am writing it in this exercise book, and just broke off for a minute to write to you.

Thank you for the address. I can't go to Paris before the spring, so I think it would be better if I did not write until then. I feel this light treatment is the right one. Not that I am ill at present. I am not in the least an invalid, in any way.

It's a sunny, windy day — beautiful. There is a soft roaring in the trees and little birds fly up into the air just for the fun of being tossed about.

Good-bye. I press your hand. Do you dislike the idea that we should write to each other from time to time?   KATHERINE.

[At the end of the manuscript of *The Garden Party.*]

This is a moderately successful story, and that's all. It's somehow, in the episode at the lane, scamped.

The New Baby

It is a late night, very dark, very still. Not a star to be seen. And now it has come on to rain. What happiness it is to listen to rain at night; joyful relief, ease; a lapping-round and hushing and brooding tenderness, all are mingled together in the sound of the fast-falling rain. God, looking down upon the rainy earth, sees how faint are these lights shining in the little windows, — how easily put out. . . .

Suddenly, quick hard steps mount the stone staircase. Someone is hurrying. There is a knock at my door, and at the same moment a red beaming face is thrust in, as Ernestine announces, "He is born."

Born!

"He is born!"

Oh, Ernestine, don't turn away. Don't be afraid. Let me weep too.

You ought to keep this, my girl, just as a *warning* to show what an arch-wallower you *can* be.

October 16. Another radiant day. J. is typing my last story, *The Garden Party*, which I finished on my birthday [October 14]. It took me nearly a month to 'recover' from *At the Bay*. I made at least three false starts. But I could not get away from the sound of the sea, and Beryl fanning her hair at the window. These things would not *die down*. But now I'm not at all sure about that story. It seems to me it's a little 'wispy' — not what it might have been. The *G.P.* is better. But that is not *good enough*, either. . . .

The last few days what one notices more than anything is the blue. Blue sky, blue mountains, all is a heavenly blueness! And clouds of all kinds — wings, soft white clouds, almost hard little golden islands, great mock-mountains. The gold deepens on the slopes. In fact, in sober fact, it is perfection.

But the late evening is the time — of times. Then with that unearthly beauty before one it is not hard to realise how far one has to go. To write something that will be worthy of that rising moon, that pale light. To be 'simple' enough, as one would be simple before God. . . .

October 27. Stories for my new book.

N.Z. *Honesty:* The Doctor, Arnold Cullen, and his wife Lydia, and Archie the friend.

L. *Second Violin:* Alexander and his friend in the train. Spring — spouting rain. *Wet lilac.*

N.Z. *Six Years After:* A wife and husband on board a steamer. The cold buttons. They see someone who reminds them.

L. *Life like Logs of Driftwood:* This wants to be a long, very well-written story. The men are important, especially the lesser man. It wants a good deal of working . . . newspaper office.

N.Z. *A Weak Heart:* Ronnie on his bike in the evening, with his hands in his pockets, *doing marvels*, by that dark tree at the corner of May Street. Edie and Ronnie.

L. *Widowed:* Geraldine and Jimmie: a house overlooking Sloane Street and Square. Wearing those buds at her heart. "Married or not married. . . ." From Autumn to Spring.

N.Z. *Our Maude:* Husband and wife play duets aňd ă ōne ă two ă thrēe ă ōne ă two thrēe ōne! His white waistcoats. Wifeling and Mahub! What a girl you are!

N.Z. *At Karori:* The little lamp. I seen it. And then they were silent. (*Finito:* October 30, 1921.)

I wish that *my* silence were only a two-minute one!

October. I wonder why it should be so very difficult to be humble. I do not think I am a good writer; I realize my faults better than anyone else could realize them. I know exactly where I fail. And yet, when I have finished a story and before I have begun another, I catch myself *preening* my feathers. It is disheartening. There seems to be some bad old pride in my heart; a root of it that puts out a thick shoot on the slightest provocation. . . . This interferes very much with work. One can't be calm, clear, good as one must be, while it goes on. I look at the mountains, I try to pray and I think of something *clever.* It's a kind of excitement within, which shouldn't be there. Calm yourself. Clear yourself. And anything that I write in this mood will be no good; it will be full of *sediment.* If I were well, I would go off by myself somewhere and sit under a tree. One must learn, one must practise, to *forget* oneself. I can't tell the truth about Aunt Anne unless I am free to look into her life without self-consciousness. Oh God! I am divided still. I am bad. I fail in my personal life. I lapse into impatience, temper, vanity, and so I fail as thy priest. Perhaps poetry will help.

I have just thoroughly cleaned and attended to my fountain pen. If after this it leaks, then it is *no* gentleman!

November 13. It is time I started a new journal. Come, my unseen, my unknown, let us talk together. Yes, for the last two weeks I have written scarcely anything. I have been idle; I have *failed.* Why? Many reasons. There has been a kind of confusion in my consciousness. It has seemed as though there was no time to write. The mornings, if they are sunny, are taken up with sun-treatment; the post eats away the afternoon. And at night I am tired.

'But it all goes deeper.' Yes, you are right. I haven't been able to yield to the kind of contemplation that is necessary. I have not felt pure in heart, not humble, not good. There's been a stirring-up of sediment. I look at the mountains and I see nothing but mountains. Be frank! I read rubbish. I give way about writing letters. I mean I refuse to meet my obligations, and this of course weakens me in every way. Then I have broken my promise to review the books for *The Nation.* Another *bad spot.* Out of hand? Yes, that describes it — dissipated, vague, not *positive* and above all, above everything, not working as I should be working — wasting time.

Wasting time. The old cry — the first and last cry — Why do ye tarry? Ah, why indeed? My deepest desire is to be a writer, to have "a body of work" done. And there the work is, there the stories wait for me, *grow tired*, wilt, fade, because I will not come. And I hear and I *acknowledge* them, and still I go sitting at the window, playing with the ball of wool. What is to be done?

I must make another effort — at once. I must begin all over again. I must try and write simply, fully, freely, from my heart. *Quietly*, caring nothing for success or failure, but just going on.

I must keep this book so that I have a record of what I do each week. (Here a word. As I re-read *At the Bay* in proof, it seemed to me flat, dull, and not a success at all. I was very much ashamed of it. I am.) But now to resolve! And especially to keep in touch with Life — with the sky and this moon, these stars, these cold, candid peaks.

November 16. To go to Sierre, if it goes on like this . . . or to — or to —

November 21. Since then [*i.e.* since writing the entry of October 16, 1921] I have only written *The Doll's House*. A bad spell has been on me. I have begun two stories,* but then I told them and they felt betrayed. It is absolutely fatal to give way to this temptation. . . . To-day I began to write, seriously, *The Weak Heart*, — a story which fascinates me *deeply*. What I feel it needs so peculiarly is a very subtle variation of 'tense' from the present to the past and back again — and softness, lightness, and the feeling that all is in bud, with a play of humour over the character of Ronnie. And the feeling of the Thorndon Baths, the wet, moist, oozy . . . no, I know how it must be done.

May I be found worthy to do it! Lord, make me crystal clear for thy light to shine through!

November 24. These last days I have been awfully rebellious. Longing for something. I feel uprooted. I want things that J. can so easily do without, that aren't natural to him. I long for them. But then, stronger than all these desires, is the other, which is to *make good* before I do anything else. The sooner the books are written, the sooner I shall be well, the sooner my wishes will be in sight of fulfilment. That is sober truth, of course. As a pure matter of fact I consider this enforced confinement here as God-given. But, on the other hand, I must make the most of it quickly. It is not unlimited any more than anything else is. Oh, why — oh, why isn't anything unlimited? Why am I troubled every single day of my life by the nearness of death and its inevitability? I am really diseased on that point. And I can't speak of it. If I tell J. it makes him unhappy. If I don't tell him, it leaves me to fight it. I am tired of the battle. No one knows how tired.

To-night, when the evening-star shone through the side-window and the

---

* Fragments of these two stories, *Widowed* and *Second Violin*, and of *Weak Heart*, are in the "Doves' Nest."

mountains were so lovely, I sat there thinking of death. Of all there was to do — of Life, which is so lovely — and of the fact that my body is a prison. But this state of mind is *evil*. It is only by acknowledging that I, being what I am, had to suffer *this* in order to do the work I am here to perform. It is only by acknowledging it, by being thankful that work was not taken away from me, that I shall recover. I am weak where I must be strong.

And to-day — Saturday — less than ever. But no matter. I have progressed . . . a little. I have realised *what* it is to be done — the strange barrier to be crossed from thinking it to writing it. . . . Daphne.

[On the next page begins the unfinished MS. of *Daphne*, included in "The Doves' Nest."]

Words

Out of us all
That make rhymes,
Will you choose
Sometimes —
As the winds use
A crack in a wall
Or a drain,
Their joy or their pain
To whistle through —
Choose me,
You English words?

I know you:
You are as light as dreams,
Tough as oak,
Precious as gold,
As poppies and corn,
Or an old cloak:
Sweet as our birds
To the ear,
As the burnet rose
In the heat
Of Midsummer:
Strange as the races
Of dead and unborn:
Strange and sweet
Equally,
And familiar,
To the eye,
As the dearest faces
That a man knows,
And as lost homes are:
But though older far
Than oldest yew, —
As our hills are, old, —
Worn new
Again and again;
Young as our streams
After rain:
And as dear

As the earth which you prove
That we love.

Make me content
With some sweetness
From Wales
Whose nightingales
Have no wings, —
From Wiltshire and Kent
And Herefordshire,
And the villages there, —
From the names, and the things
No less.

Let me sometimes dance
With you,
Or climb
Or stand perchance
In ecstasy,
Fixed and free
In a rhyme,
As poets do.

Edward Thomas (1878–1917)

GEORGE ORWELL

George Orwell was the pen-name of Eric Blair (1903–50). Educated at Eton, he served in the Indian Imperial Police before fighting in the Spanish Civil War. A series of novels and the two powerful political satires *Animal Farm and Nineteen Eighty Four* testify to his hatred of totalitarianism and sympathy for the oppressed — concerns emerging in this essay.

Politics and the English Language

Most people who bother with the matter at all would admit that the English language is in a bad way, but it is generally assumed that we cannot by conscious action do anything about it. Our civilization is decadent and our language — so the argument runs — must inevitably share in the general collapse. It follows that any struggle against the abuse of language is a sentimental archaism, like preferring candles to electric light or hansom cabs to aeroplanes. Underneath this lies the half-conscious belief that language is a natural growth and not an instrument which we shape for our own purposes.

Now it is clear that the decline of a language must ultimately have political and economic causes: it is not due simply to the bad influence of this or that individual writer. But an effect can become a cause, reinforcing the original cause and producing the same effect in an intensified form, and so on indefinitely. A man may take to drink because he feels himself a failure, and then fail all the more completely because he drinks. It is rather the same thing that is happening to the English language. It becomes ugly and inaccurate because our thoughts are foolish, but the slovenliness of our language makes it easier for us to have foolish thoughts. The point is that the process is reversible. Modern English, especially written English, is full of bad habits which spread by imitation and which can be avoided if one is willing to take the necessary trouble. If one gets rid of these habits one can think more clearly, and to think clearly is a necessary first step towards political regeneration: so that the fight against bad English is not frivolous and is not the exclusive concern of professional writers. I will come back to this presently, and I hope that by that time the meaning of what I have said here will have become clearer. Meanwhile, here are five specimens of the English language as it is now habitually written.

These five passages have not been picked out because they are especially bad — I could have quoted far worse if I had chosen — but because they illustrate various of the mental vices from which we now suffer. They are a little below the average, but are fairly representative samples. I number them so that I can refer back to them when necessary:

(1) I am not, indeed, sure whether it is not true to say that the Milton who once seemed not unlike a seventeenth-century Shelley had not become, out of an experience ever more bitter in each year, more alien [*sic*] to the founder of that Jesuit sect which nothing could induce him to tolerate.

Professor Harold Laski (Essay in *Freedom of Expression*)

(2) Above all, we cannot play ducks and drakes with a native battery of idioms which prescribes such egregious collocations of vocables as the Basic *put up with* for *tolerate* or *put at a loss* for *bewilder*.

Professor Lancelot Hogben (*Interglossa*)

(3) On the one side we have the free personality: by definition it is not neurotic, for it has neither conflict nor dream. Its desires, such as they are, are transparent, for they are just what institutional approval keeps in the forefront of consciousness; another institutional pattern would alter their number and intensity; there is little in them that is natural, irreducible, or culturally dangerous. But *on the other side*, the social bond itself is nothing but the mutual reflection of these self-secure integrities. Recall the definition of love. Is not this the very picture of a small academic? Where is there a place in this hall of mirrors for either personality or fraternity?

Essay on psychology in *Politics* (New York)

(4) All the 'best people' from the gentlemen's clubs, all the frantic fascist captains, united in common hatred of Socialism and bestial horror of the rising tide of the mass revolutionary movement, have turned to acts of provocation, to foul incendiarism, to medieval legends of poisoned wells, to legalize their own destruction of proletarian organizations, and rouse the agitated petty-bourgeoisie to chauvinistic fervour on behalf of the fight against the revolutionary way out of the crisis.

Communist pamphlet

(5) If a new spirit *is* to be infused into this old country, there is one thorny and contentious reform which must be tackled, and that is the humanization and galvanization of the B.B.C. Timidity here will bespeak canker and atrophy of the soul. The heart of Britain may be sound and of a strong beat, for instance, but the British lion's roar at present is like that of Bottom in Shakespeare's *Midsummer Night's Dream* — as gentle as any sucking dove. A virile new Britain cannot continue indefinitely to be traduced in the eyes or rather ears, of the world by the effete languors of Langham Place, brazenly masquerading as 'standard English.' When the Voice of Britain is heard at nine o'clock, better far and infinitely less ludicrous to hear aitches honestly dropped than the present priggish, inflated, inhibited, school-ma'amish arch braying of blameless bashful mewing maidens!

Letter in *Tribune*

Each of these passages has faults of its own, but, quite apart from avoidable ugliness, two qualities are common to all of them. The first is staleness of imagery; the other is lack of precision. The writer either has a meaning and cannot express it, or he inadvertently says something else, or he is almost indifferent as to whether his words mean anything or not. This mixture of vagueness and sheer incompetence is the most marked characteristic of modern English prose, and especially of any kind of political writing. As soon as certain topics are raised, the concrete melts into the abstract and no one seems able to think of turns of speech that are not hackneyed: prose consists less and less of *words* chosen for the sake of their meaning, and more and more of *phrases* tacked together like the sections of a prefabricated hen-house. I list below, with notes and examples, various of the tricks by means of which the work of prose-construction is habitually dodged:

DYING METAPHORS. A newly invented metaphor assists thought by evoking a visual image, while on the other hand a metaphor which is technically 'dead' (e.g. *iron resolution*) has in effect reverted to being an ordinary word and can generally be used without loss of vividness. But in between these two classes there is a huge dump of worn-out metaphors which have lost all evocative power and are merely used because they save people the trouble of inventing phrases for themselves. Examples are: *Ring the changes on, take up the cudgels for, toe the line, ride roughshod over, stand shoulder to shoulder with, play into the hands of, no axe to grind, grist to the mill, fishing in troubled waters, on the order of the day, Achilles' heel, swan song, hotbed*. Many of these are used without knowledge of their meaning (what is a 'rift,' for instance?), and incompatible metaphors are frequently mixed, a sure sign that the writer is not interested in what he is saying. Some metaphors now current have been twisted out of their original meaning without those who use them even being aware of the fact. For example, *toe the line* is sometimes written *tow the line*. Another example is *the hammer and the anvil*, now always used with the implication that the anvil gets the worst of it. In real life it is always the anvil that breaks the hammer, never the other way about: a writer who stopped to think what he was saying would be aware of this, and would avoid perverting the original phrase.

OPERATORS or VERBAL FALSE LIMBS. These save the trouble of picking out appropriate verbs and nouns, and at the same time pad each sentence with extra syllables which give it an appearance of symmetry. Characteristic phrases are *render inoperative, militate against, make contact with, be subjected to, give rise to, give grounds for, have the effect of, play a leading part (role) in, make itself felt, take effect, exhibit a tendency to, serve the purpose of, etc., etc.* The keynote is the elimination of simple verbs. Instead of being a single word, such as *break, stop, spoil, mend, kill*, a verb becomes a *phrase*, made up of a noun or adjective tacked on to some general-purposes verb such as *prove, serve, form, play, render* . In addition, the passive voice is wherever possible used in preference to the active, and noun constructions are used instead of gerunds (*by examination of* instead of *by examining*). The range of verbs is further cut down by means of the *-ize* and *de-* formations, and the banal statements are given an appearance of profundity by means of the *not un-* formation. Simple conjunctions and prepositions are replaced by such phrases as *with respect to, having regard to, the fact that, by dint of, in view of, in the interests of, on the hypothesis that;* and the ends of sentences are saved from anticlimax by such resounding commonplaces as *greatly to be desired, cannot be left out of account, a development to be expected in the near future, deserving of serious consideration, brought to a satisfactory conclusion*, and so on and so forth.

PRETENTIOUS DICTION. Words like *phenomenon, element, individual* (as noun), *objective, categorical, effective, virtual, basic, primary, promote, constitute, exhibit, exploit, utilize, eliminate, liquidate*, are used to dress up simple statements and give an air of scientific impartiality to biased judgements. Adjectives like *epoch-making, epic, historic, unforgettable, triumphant, age-old, inevitable, inexorable, veritable* are used to dignify the sordid processes of international politics, while writing that aims at glorifying war usually takes on an archaic colour, its characteristic words being: *realm, throne, chariot, mailed fist, trident, sword, shield, buckler, banner, jackboot, clarion*. Foreign words and expressions such as *cul de sac, ancien régime, deus ex machina, mutatis mutandis, status quo, gleichschaltung, weltanschauung*, are used to give an air of culture and elegance. Except for the useful abbreviations *i.e., e.g.,* and *etc.,* there is no real need for any of the hundreds of foreign phrases now current in English. Bad writers, and especially scientific, political, and sociological writers, are nearly always haunted by the notion that Latin or Greek words are grander than Saxon ones, and unnecessary words like *expedite, ameliorate, predict, extraneous, deracinated, clandestine, subaqueous,* and hundred of others constantly gain ground from their Anglo-Saxon opposite numbers.* The jargon peculiar to Marxist writing (*hyena, hangman, cannibal, petty bourgeois, these gentry, lackey, flunkey, mad dog, White Guard*, etc.) consists largely of words and phrases translated from Russian, German or French; but the normal way of coining a new word is to use a Latin or Greek root with the appropriate affix and, where necessary, the -ize formation. It is often easier to make up words of

---

* An interesting illustration of this is the way in which the English flower names which were in use till very recently are being ousted by Greek ones, *snapdragon* becoming *antirrhinum, forget-me-not* becoming *myosotis*, etc. It is hard to see any practical reason for this change of fashion; it is probably due to an instinctive turning away from the more homely word and a vague feeling that the Greek word is scientific.

this kind (*deregionalize, impermissible, extramarital, non-fragmentary* and so forth) than to think up the English words that will cover one's meaning. The result, in general, is an increase in slovenliness and vagueness.

MEANINGLESS WORDS. In certain kinds of writing, particularly in art criticism and literary criticism, it is normal to come across long passages which are almost completely lacking in meaning.* Words like *romantic, plastic, values, human, dead, sentimental, natural, vitality,* as used in art criticism, are strictly meaningless, in the sense that they not only do not point to any discoverable object, but are hardly ever expected to do so by the reader. When one critic writes, 'The outstanding feature of Mr. X's work is its living quality,' while another writes, 'The immediately striking thing about Mr. X's work is its peculiar deadness,' the reader accepts this as a simple difference of opinion. If words like *black* and *white* were involved, instead of the jargon words *dead* and *living*, he would see at once that language was being used in an improper way. Many political words are similarly abused. The word *Fascism* has now no meaning except in so far as it signifies 'something not desirable.' The words *democracy, socialism, freedom, patriotic, realistic, justice,* have each of them several different meanings which cannot be reconciled with one another. In the case of a word like *democracy*, not only is there no agreed definition, but the attempt to make one is resisted from all sides. It is almost universally felt that when we call a country democratic we are praising it; consequently the defenders of every kind of régime claim that it is a democracy, and fear that they might have to stop using the word if it were tied down to any one meaning. Words of this kind are often used in a consciously dishonest way. That is, the person who uses them has his own private definition, but allows his hearer to think he means something quite different. Statements like *Marshal Pétain was a true patriot, The Soviet press is the freest in the world, The Catholic Church is opposed to persecution*, are almost always made with intent to deceive. Other words used in variable meanings, in most cases more or less dishonestly, are: *class, totalitarian, science, progressive, reactionary, bourgeois, equality.*

Now that I have made this catalogue of swindles and perversions, let me give another example of the kind of writing that they lead to. This time it must of its nature be an imaginary one. I am going to translate a passage of good English into modern English of the worst sort. Here is a well-known verse from *Ecclesiastes:*

---

*Example: 'Comfort's catholicity of perception and image, strangely Whitmanesque in range, almost the exact opposite in aesthetic compulsion, continues to evoke that trembling atmospheric accumulative hinting at a cruel, an inexorably sereve timelessness . . . Wrey Gardiner scores by aiming at simple bull's-eyes with precision. Only they are not so simple, and through this contented sadness runs more than the surface bittersweet of resignation. (*Poetry Quarterly.*)

> I returned and saw under the sun, that the race is not to the swift, nor the battle to the strong, neither yet bread to the wise, nor yet riches to men of understanding, nor yet favour to men of skill; but time and chance happeneth to them all.

Here it is in modern English:

> Objective consideration of contemporary phenomena compels the conclusion that success or failure in competitive activities exhibits no tendency to be commensurate with innate capacity, but that a considerable element of the unpredictable must invariably be taken into account.

This is a parody, but not a very gross one. Exhibit (3), above, for instance, contains several patches of the same kind of English. It will be seen that I have not made a full translation. The beginning and the ending of the sentence follow the original meaning fairly closely, but in the middle the concrete illustrations — race, battle, bread — dissolve into the vague phrase 'success or failure in competitive activities.' This had to be so, because no modern writer of the kind I am discussing — no one capable of using phrases like 'objective consideration of contemporary phenomena' — would ever tabulate his thoughts in that precise and detailed way. The whole tendency of modern prose is away from concreteness. Now analyze these two sentences a little more closely. The first contains forty-nine words but only sixty syllables, and all its words are those of everyday life. The second contains thirty-eight words of ninety syllables: eighteen of its words are from Latin roots, and one from Greek. The first sentence contains six vivid images, and only one phrase ('time and chance') that could be called vague. The second contains not a single fresh, arresting phrase, and in spite of its ninety syllables it gives only a shortened version of the meaning contained in the first. Yet without a doubt it is the second kind of sentence that is gaining ground in modern English. I do not want to exaggerate. This kind of writing is not yet universal, and outcrops of simplicity will occur here and there in the worst-written page. Still, if you or I were told to write a few lines on the uncertainty of human fortunes, we should probably come much nearer to my imaginary sentence than to the one from *Ecclesiastes*.

As I have tried to show, modern writing at its worst does not consist in picking out words for the sake of their meaning and inventing images in order to make the meaning clearer. It consists in gumming together long strips of words which have already been set in order by someone else, and making the results presentable by sheer humbug. The attraction of this way of writing is that it is easy. It is easier — even quicker, once you have the habit — to say *In my opinion it is not an unjustifiable assumption that* than to say *I think*. If you use ready-made phrases, you not only don't have to hunt about for words; you also don't have to bother with the rhythms of your sentences, since these phrases are generally so arranged as to be more or less euphonious. When you are composing in a hurry — when you are dictating to a stenographer, for instance, or making a public speech — it is natural to fall into a pretentious, Latinized style. Tags like *a consideration which we should do well to bear in mind* or *a conclusion to which all of us would readily assent* will save many a sentence from coming down with a bump. By using stale metaphors, similes and idioms, you save much mental effort, at the cost of leaving your meaning vague, not only for your reader but for yourself. This is the significance of mixed metaphors. The sole aim of a metaphor is to call up a visual image. When these images clash — as in *The Fascist octopus has sung its swan song, the jackboot is thrown into the melting pot* — it can be taken as certain that the writer is not seeing a mental image of the objects he is naming; in other words he is not really thinking. Look again at the examples I gave at the beginning of this essay. Professor Laski (1) uses five negatives in fifty-three words. One of these is superfluous, making nonsense of the whole passage, and in addition there is the slip *alien* for akin, making further nonsense, and several avoidable pieces of clumsiness which increase the general vagueness. Professor Hogben (2) plays ducks and drakes with a battery which is able to write prescriptions, and, while disapproving of the everyday phrase *put up with*, is unwilling to look *egregious* up in the dictionary and see what it means. (3), if one takes an uncharitable attitude towards it, is simply meaningless: probably one could work out its intended meaning by reading the whole of the article in which it occurs. In (4), the writer knows more or less what he wants to say, but an accumulation of stale phrases chokes him like tea leaves blocking a sink. In (5), words and meaning have almost parted company. People who write in this manner usually have a general emotional meaning — they dislike one thing and want to express solidarity with another — but they are not interested in the detail of what they are saying. A scrupulous writer, in every sentence that he writes, will ask himself at least four questions, thus: What am I trying to say? What words will express it? What image or idiom will make it clearer? Is this image fresh enough to have an effect? And he will probably ask himself two more: Could I put it more shortly? Have I said anything that is avoidably ugly? But you are not obliged to go to all this trouble. You can shirk it by simply throwing your mind open and letting the ready-made phrases come crowding in. They will construct your sentences for you — even think your thoughts for you, to a certain extent — and at need they will perform the important service of partially concealing your meaning even from yourself. It is at this point that the special connection between politics and the debasement of language becomes clear.

In our time it is broadly true that political writing is bad writing. Where it is not true, it will generally be found that the writer is some kind of rebel, expressing his private opinions and not a 'party line.' Orthodoxy, of whatever colour, seems to demand a lifeless, imitative style. The political dialects to be found in pamphlets, leading articles, manifestos, White Papers and the speeches of undersecretaries do, of course, vary from party to party, but they are all alike in that one almost never finds in them a fresh, vivid, home-made turn of speech. When one watches some tired hack on the platform mechanically repeating the familiar phrases — *bestial atrocities, iron heel, bloodstained tyranny, free peoples of the world, stand shoulder to shoulder* — one often has a curious feeling that one is not watching a live human being but some kind of dummy: a feeling which suddenly becomes stronger at moments when the light catches the speaker's spectacles and turns them into blank discs which seem to have no eyes behind them. And this is not altogether fanciful. A speaker who uses that kind of phraseology has gone some distance towards turning himself into a machine. The appropriate noises are coming out of his larynx, but his brain is not involved as it would be if he were choosing his words for himself. If the speech he is making is one that he is accustomed to make over and over again, he may be almost unconscious of what he is saying, as one is when one utters the responses in church. And this reduced state of consciousness, if not indispensable, is at any rate favourable to political conformity.

In our time, political speech and writing are largely the defence of the indefensible. Things like the continuance of British rule in India, the Russian purges and deportations, the dropping of the atom bombs on Japan, can indeed be defended, but only by arguments which are too brutal for most people to face, and which do not square with the professed aims of political parties. Thus political language has to consist largely of euphemism, question-begging and sheer cloudy vagueness. Defenceless villages are bombarded from the air, the inhabitants driven out into the countryside, the cattle machine-gunned, the huts set on fire with incendiary bullets: this is called *pacification*. Millions of peasants are robbed of their farms and sent trudging along the roads with no more than they can carry: this is called *transfer of population* or *rectification of frontiers*. People are imprisoned for years without trial, or shot in the back of the neck or sent to die of scurvy in Arctic lumber camps: this is called *elimination of unreliable elements*. Such phraseology is needed if one wants to name things without calling up mental pictures of them. Consider for instance some comfortable English professor defending Russian totalitarianism. He cannot say outright, 'I believe in killing off your opponents when you can get good results by doing so.' Probably, therefore, he will say something like this:

'While freely conceding that the Soviet régime exhibits certain features which the humanitarian may be inclined to deplore, we must, I think, agree that a certain curtailment of the right to political opposition is an unavoidable concomitant of transitional periods, and that the rigours which the Russian people have been called upon to undergo have been amply justified in the sphere of concrete achievement.'

The inflated style is itself a kind of euphemism. A mass of Latin words falls upon the facts like soft snow, blurring the outlines and covering up all the details. The great enemy of clear language is insincerity. When there is a gap between one's real and one's declared aims, one turns as it were instinctively to long words and exhausted idioms, like a cuttlefish squirting out ink. In our age there is no such thing as 'keeping out of politics.' All issues are political issues, and politics itself is a mass of lies, evasions, folly, hatred and schizophrenia. When the general atmosphere is bad, language must suffer. I should expect to find — this is a guess which I have not sufficient knowledge to verify — that the German, Russian and Italian languages have all deteriorated in the last ten or fifteen years, as a result of dictatorship.

But if thought corrupts language, language can also corrupt thought. A bad usage can spread by tradition and imitation, even among people who should and do know better. The debased language that I have been discussing is in some ways very convenient. Phrases like *a not unjustifiable assumption, leaves much to be desired, would serve no good purpose, a consideration which we should do well to bear in mind*, are a continuous temptation, a packet of aspirins always at one's elbow. Look back through this essay, and for certain you will find that I have again and again committed the very faults I am protesting against. By this morning's post I have received a pamphlet dealing with conditions in Germany. The author tells me that he 'felt impelled' to write it. I open it at random, and here is almost the first sentence that I see: '(The Allies) have an opportunity not only of achieving a radical transformation of Germany's social and political structure in such a way as to avoid a nationalistic reaction in Germany itself, but at the same time of laying the foundations of a co-operative and unified Europe.' You see, he 'feels impelled' to write — feels, presumably, that he has something new to say — and yet his words, like cavalry horses answering the bugle, group themselves automatically into the familiar dreary pattern. This invasion of one's mind, by ready-made phrases (*lay the foundation, achieve a radical transformation*) can only be prevented if one is constantly on guard against them, and every such phrase anaesthetizes a portion of one's brain.

I said earlier that the decadence of our language is probably curable. Those who deny this would argue, if they produced an argument at all, that language merely reflects existing social conditions, and that we cannot influence its development by any direct tinkering with words and constructions. So far as the general tone or spirit of a language goes, this may be true, but it is not true in detail. Silly words and expressions have often disappeared, not through any evolutionary process but owing to the conscious action of a minority. Two recent examples were *explore every avenue* and *leave no stone unturned*, which were killed by the jeers of a few journalists. There is a long list of flyblown metaphors which could similarly be got rid of if enough people would interest themselves in the job; and it should also be possible to laugh the *not un-* formation out of existence,* to reduce the amount of Latin and Greek in the average sentence, to drive out foreign phrases and strayed scientific words, and, in general, to make pretentiousness unfashionable. But these are all minor points. The defence of the English language implies more than this, and perhaps it is best to start by saying what it *does not* imply.

To begin with it has nothing to do with archaism, with the salvaging of obsolete words and turns of speech, or with the setting up of a 'standard English' which must never be departed from. On the contrary, it is especially concerned with the scrapping of every word or idiom which has outworn its usefulness. It has nothing to do with correct grammar and syntax, which are of no importance so long as one makes one's meaning clear, or with the avoidance of Americanisms, or with having what is called a 'good prose style.' On the other hand it is not concerned with fake simplicity and the attempt to make written English colloquial. Nor does it even imply in every case preferring the Saxon word to the Latin one, though it does imply using the fewest and shortest words that will cover one's meaning. What is above all needed is to let the meaning choose the word, and not the other way about. In prose, the worst thing one can do with words is to surrender to them. When you think of a concrete object, you think wordlessly, and then, if you want to describe the thing you have been visualizing you probably hunt about until you find the exact words that seem to fit it. When you think of something abstract you are more inclined to use words from the start, and unless you make a conscious effort to prevent it, the existing dialect will come rushing in and do the job for you, at the expense of blurring or even changing your meaning. Probably it is better to put off using words as long as possible and get one's meaning as clear as one can through pictures or sensations. Afterward one can choose — not simply *accept* — the phrases that will best cover the meaning, and then switch round and decide what impression one's words are likely to make on another person. This last effort of the mind cuts out all stale or mixed images, all prefabricated phrases, needless repetitions, and humbug and vagueness generally. But one can often be in doubt about the effect of a word or a phrase, and one needs rules that one can rely on when instinct fails. I think the following rules will cover most cases:

(i) Never use a metaphor, simile, or other figure of speech which you are used to seeing in print.

(ii) Never use a long word where a short one will do.

(iii) If it is possible to cut out a word, always cut it out.

(iv) Never use the passive where you can use the active.

(v) Never use a foreign phrase, a scientific word or a jargon word if you can think of an everyday English equivalent.

---

* One can cure oneself of the *not un-* formation by memorizing this sentence: *A not unblack dog was chasing a not unsmall rabbit across a not ungreen field.*

(vi) Break any of these rules sooner than say anything outright barbarous.

These rules sound elementary, and so they are, but they demand a deep change of attitude in anyone who has grown used to writing in the style now fashionable. One could keep all of them and still write bad English, but one could not write the kind of stuff that I quoted in those five specimens at the beginning of this article.

I have not been considering the literary use of language, but merely language as an instrument for expressing and not for concealing or preventing thought. Stuart Chase and others have come near to claiming that all abstract words are meaningless, and have used this as a pretext for advocating a kind of political quietism. Since you don't know what Fascism is, how can you struggle against Fascism? One need not swallow such absurdities as this, but one ought to recognize that the present political chaos is concerned with the decay of language, and that one can probably bring about some improvement by starting at the verbal end. If you simplify your English, you are freed from the worst follies of orthodoxy. You cannot speak any of the necessary dialects, and when you make a stupid remark its stupidity will be obvious, even to yourself. Political language — and with variations this is true of all political parties, from Conservatives to Anarchists — is designed to make lies sound truthful and murder respectable, and to give an appearance of solidity to pure wind. One cannot change this all in a moment, but one can at least change one's own habits, and from time to time one can even, if one jeers loudly enough, send some worn-out and useless phrase — some *jackboot*, *Achilles' heel*, *hotbed*, *melting pot*, *acid test*, *veritable inferno*, or other lump of verbal refuse — into the dustbin where it belongs.

since feeling is first

since feeling is first
who pays any attention
to the syntax of things
will never wholly kiss you;

wholly to be a fool
while Spring is in the world

my blood approves,
and kisses are a better fate
than wisdom
lady i swear by all flowers. Don't cry
— the best gesture of my brain is less than
your eyelids' flutter which says

we are for each other: then
laugh, leaning back in my arms
for life's not a paragraph

And death i think is no parenthesis

e. e. cummings (1894–1962)

GEORGE ORWELL

A characteristically and relentlessly honest piece of Orwellian self-analysis, with some perceptive insight into the force and fallibility of language.

Why I Write

From a very early age, perhaps the age of five or six, I knew that when I grew up I should be a writer. Between the ages of about seventeen and twenty-four I tried to abandon this idea, but I did so with the consciousness that I was outraging my true nature and that sooner or later I should have to settle down and write books.

I was the middle child of three, but there was a gap of five years on either side, and I barely saw my father before I was eight. For this and other reasons I was somewhat lonely, and I soon developed disagreeable mannerisms which made me unpopular throughout my schooldays. I had the lonely child's habit of making up stories and holding conversations with imaginary persons, and I think from the very start my literary ambitions were mixed up with the feeling of being isolated and undervalued. I knew that I had a facility with words and a power of facing unpleasant facts, and I felt that this created a sort of private world in which I could get my own back for my failure in everyday life. Nevertheless the volume of serious — i.e. seriously intended — writing which I produced all through my childhood and boyhood would not amount to half a dozen pages. I wrote my first poem at the age of four or five, my mother taking it down to dictation. I cannot remember anything about it except that it was about a tiger and the tiger had 'chair-like teeth' — a good enough phrase, but I fancy the poem was a plagiarism of Blake's 'Tiger, Tiger.' At eleven, when the war of 1914-18 broke out, I wrote a patriotic poem which was printed in the local newspaper, as was another, two years later, on the death of Kitchener.[1] From time to time, when I was a bit older, I wrote bad and usually unfinished 'nature poems' in the Georgian style. I also, about twice, attempted a short story which was a ghastly failure. That was the total of the would-be serious work that I actually set down on paper during all those years.

However, throughout this time I did in a sense engage in literary activities. To begin with there was the made-to-order stuff which I produced quickly, easily and without much pleasure to myself. Apart from school work, I wrote *vers d'occasion*,[2] semi-comic poems which I could turn out at what now seems to me astonishing speed — at fourteen I wrote a whole rhyming play, in imitation of Aristophanes, in about a week — and helped to edit school magazines, both printed and in manuscript. These magazines were the most pitiful burlesque stuff that you could imagine, and I took far less trouble with them than I now would with the cheapest journalism. But side by side with all this, for fifteen years or more, I was carrying out a literary exercise of a quite different kind: this was the making up of a continuous 'story' about myself, a sort of diary existing only in the mind. I believe this is a common habit of children and adolescents. As a very small child I used to imagine that I was, say, Robin Hood, and picture myself as the hero of thrilling adventures, but quite soon my 'story' ceased to be narcissistic in a crude way and became more and more a mere description of what I was doing and the things I saw. For minutes at a time this kind of thing would be running through my head: 'He pushed the door open and entered the room. A yellow beam of sunlight, filtering through the muslin curtains, slanted on to the table, where a matchbox, half open, lay beside the inkpot. With his right hand in his pocket he moved across to the window. Down in the street a tortoiseshell cat was chasing a dead leaf,' etc., etc. This habit continued till I was about twenty-five, right through my non-literary years. Although I had to search, and did search, for the right words, I seemed to be making this descriptive effort almost against my will, under a kind of compulsion from outside. The 'story' must, I suppose, have reflected the styles of the various writers I admired at different ages, but so far as I remember it always had the same meticulous descriptive quality.

When I was about sixteen I suddenly discovered the joy of mere words, *i.e.* the sounds and associations of words. The lines from *Paradise Lost* —

> So hee with difficulty and labour hard
> Moved on: with difficulty and labour hee,[3]

which do not now seem to me so very wonderful, sent shivers down my backbone; and the spelling 'hee' for 'he' was an added pleasure. As for the need to describe things, I knew about it already. So it is clear what kind of books I wanted to write, in so far as I could be said to want to write books at that time. I wanted to write enormous naturalistic novels with unhappy endings, full of detailed descriptions and arresting similes, and also full of purple passages in which words were used partly for the sake of their sound. And in fact my first completed novel, *Burmese Days*, which I wrote when I was thirty but projected much earlier, is rather that kind of book.

I give all this background information because I do not think one can assess a writer's motives without knowing something of his early development. His subject matter will be determined by the age he lives in — at least this is true in tumultuous, revolutionary ages like our own — but before he ever begins to write he will have acquired an emotional attitude from which he will never completely escape. It is his job, no doubt, to discipline his temperament and avoid getting stuck at some immature stage, or in some perverse mood: but if he escapes from his early influences altogether, he will have killed his impulse to write. Putting aside the need to earn a living, I think there are four great motives for writing, at any rate for writing prose. They exist in different degrees in every writer, and in any one writer the proportions will vary from time to time, according to the atmosphere in which he is living. They are:

(1) Sheer egoism. Desire to seem clever, to be talked about, to be remembered after death, to get your own back on grown-ups who snubbed you in childhood, etc., etc. It is humbug to pretend that this is not a motive, and a strong one. Writers share this characteristic with scientists, artists, politicians, lawyers, soldiers, successful businessmen — in short, with the whole top crust of humanity. The great mass of human beings are not acutely selfish. After the age of about thirty they abandon individual ambition — in many cases, indeed, they almost abandon the sense of being individuals at all — and live chiefly for others, or are simply smothered under drudgery. But there is also the minority of gifted, wilful people who are determined to live their own lives to the end, and writers belong in this class. Serious writers, I should say, are on the whole more vain and self-centred than journalists, though less interested in money.

(2) Æsthetic enthusiasm. Perception of beauty in the external world, or, on the other hand, in words and their right arrangement. Pleasure in the impact of one sound on another, in the firmness of good prose or the rhythm of a good story. Desire to share an experience which one feels is valuable and ought not to be missed. The aesthetic motive is very feeble in a lot of writers, but even a pamphleteer or a writer of textbooks will have pet words and phrases which appeal to him for non-utilitarian reasons; or he may feel strongly about typography, width of margins, etc. Above the level of a railway guide, no book is quite free from æsthetic considerations.

(3) Historical impulse. Desire to see things as they are, to find out true facts and store them up for the use of posterity.

(4) Political purpose — using the word 'political' in the widest possible sense. Desire to push the world in a certain direction, to alter other people's idea of the kind of society that they should strive after. Once again, no book is genuinely free from political bias. The opinion that one should have nothing to do with politics is itself a political attitude.

It can be seen how these various impulses must war against one another, and how they must fluctuate from person to person and from time to time. By nature — taking your 'nature' to be the state you have attained when you are first adult — I am a person in whom the first three motives would outweigh the fourth. In a peaceful age I might have written ornate or merely descriptive books, and might have remained almost unaware of my political loyalties. As it is I have been forced into becoming a sort of pamphleteer.

First I spent five years in an unsuitable profession (the Indian Imperial Police, in Burma), and then I underwent poverty and the sense of failure. This increased my natural hatred of authority and made me for the first time fully aware of the existence of the working classes, and the job in Burma had given me an understanding of the nature of imperialism: but these experiences were not enough to give me an accurate political orientation. Then came Hitler, the Spanish civil war, etc. By the end of 1935 I had still failed to reach a firm decision. I remember a little poem that I wrote at that date, expressing my dilemma:

> A happy vicar I might have been
> Two hundred years ago,
> To preach upon eternal doom
> And watch my walnuts grow;
>
> But born, alas, in an evil time,
> I missed that pleasant haven,
> For the hair has grown on my upper lip
> And the clergy are all clean-shaven.
>
> And later still the times were good,
> We were so easy to please,
> We rocked our troubled thoughts to sleep
> On the bosoms of the trees.
>
> All ignorant we dared to own
> The joys we now dissemble;
> The greenfinch on the apple bough
> Could make my enemies tremble.
>
> But girls' bellies and apricots,
> Roach in a shaded stream,
> Horses, ducks in flight at dawn,
> All these are a dream.
>
> It is forbidden to dream again;
> We maim our joys or hide them;
> Horses are made of chromium steel
> And little fat men shall ride them.
>
> I am the worm who never turned,
> The eunuch without a harem;
> Between the priest and the commissar
> I walk like Eugene Aram;[4]
>
> And the commissar is telling my fortune
> While the radio plays,
> But the priest has promised an Austin Seven,
> For Duggie always pays.

I dreamed I dwelt in marble halls,[5]
And woke to find it true;
I wasn't born for an age like this;
Was Smith? Was Jones? Were you?

The Spanish war and other events in 1936–7 turned the scale and thereafter I knew where I stood. Every line of serious work that I have written since 1936 has been written, directly or indirectly, *against* totalitarianism and *for* democratic socialism, as I understand it. It seems to me nonsense, in a period like our own, to think that one can avoid writing of such subjects. Everyone writes of them in one guise or another. It is simply a question of which side one takes and what approach one follows. And the more one is conscious of one's political bias, the more chance one has of acting politically without sacrificing one's æsthetic and intellectual integrity.

What I have most wanted to do throughout the past ten years is to make political writing into an art. My starting point is always a feeling of partisanship, a sense of injustice. When I sit down to write a book, I do not say to myself, 'I am going to produce a work of art.' I write it because there is some lie that I want to expose, some fact to which I want to draw attention, and my initial concern is to get a hearing. But I could not do the work of writing a book, or even a long magazine article, if it were not also an æsthetic experience. Anyone who cares to examine my work will see that even when it is downright propaganda it contains much that a full-time politician would consider irrelevant. I am not able, and I do not want, completely to abandon the world-view that I acquired in childhood. So long as I remain alive and well I shall continue to feel strongly about prose style, to love the surface of the earth, and to take a pleasure in solid objects and scraps of useless information. It is no use trying to suppress that side of myself. The job is to reconcile my ingrained likes and dislikes with the essentially public, non-individual activities that this age forces on all of us.

It is not easy. It raises problems of construction and of language, and it raises in a new way the problem of truthfulness. Let me give just one example of the cruder kind of difficulty that arises. My book about the Spanish civil war, *Homage to Catalonia*, is of course, a frankly political book, but in the main it is written with a certain detachment and regard for form. I did try very hard in it to tell the whole truth without violating my literary instincts. But among other things it contains a long chapter, full of newspaper quotations and the like, defending the Trotskyists who were accused of plotting with Franco. Clearly such a chapter, which after a year or two would lose its interest for any ordinary reader, must ruin the book. A critic whom I respect read me a lecture about it. 'Why did you put in all that stuff?' he said. 'You've turned what might have been a good book into journalism.' What he said was true, but I could not have done otherwise. I happened to know, what very few people in England had been allowed to know, that innocent men were being falsely accused. If I had not been angry about that I should never have written the book.

In one form or another this problem comes up again. The problem of language is subtler and would take too long to discuss. I will only say that of late years I have tried to write less picturesquely and more exactly. In any case I find that by the time you have perfected any style of writing, you have always outgrown it. *Animal Farm* was the first book in which I tried, with full consciousness of what I was doing, to fuse political purpose and artistic purpose into one whole. I have not written a novel for seven years, but I hope to write another fairly soon.[6] It is bound to be a failure, every book is a failure, but I do know with some clarity what kind of book I want to write.

Looking back through the last page or two, I see that I have made it appear as though my motives in writing were wholly public-spirited. I don't want to leave that as the final impression. All writers are vain, selfish and lazy, and at the very bottom of their motives there lies a mystery. Writing a book is a horrible, exhausting struggle, like a long bout of some painful illness. One would never undertake such a thing if one were not driven on by some demon whom one can neither resist nor understand. For all one knows that demon is simply the same instinct that makes a baby squall for attention. And yet it is also true that one can write nothing readable unless one constantly struggles to efface one's own personality. Good prose is like a window pane. I cannot say with certainty which of my motives are the strongest, but I know which of them deserve to be followed. And looking back through my work, I see that it is invariably where I lacked a *political* purpose that I wrote lifeless books and was betrayed into purple passages, sentences without meaning, decorative adjectives and humbug generally.

To Juan at the Winter Solstice

There is one story and one story only
That will prove worth your telling,
Whether as learned bard or gifted child;
To it all lines or lesser gauds belong
That startle with their shining
Such common stories as they stray into.

Is it of trees you tell, their months and virtues,
Or strange beasts that beset you,
Of birds that croak at you the Triple will?
Or of the Zodiac and how slow it turns
Below the Boreal Crown,
Prison of all true kings that ever reigned?

Water to water, ark again to ark,
From woman back to woman:
So each new victim treads unfalteringly
The never altered circuit of his fate,
Bringing twelve peers as witness
Both to his starry rise and starry fall.

Or is it of the Virgin's silver beauty,
All fish below the thighs?
She in her left hand bears a leafy quince;
When, with her right she crooks a finger smiling,
How may the King hold back?
Royally then he barters life for love.

Or of the undying snake from chaos hatched,
Whose coils contain the ocean,
Into whose chops with naked sword he springs,
Then in black water, tangled by the reeds,
Battles three days and nights,
To be spewed up beside her scalloped shore?

Much snow is falling, winds roar hollowly,
The owl hoots from the elder,
Fear in your heart cries to the loving-cup:
Sorrow to sorrow as the sparks fly upward.

The log groans and confesses
There is one story and one story only.

Dwell on her graciousness, dwell on her smiling,
Do not forget what flowers
The great boar trampled down in ivy time.
Her brow was creamy as the crested wave,
Her sea-blue eyes were wild
But nothing promised that is not performed.

Robert Graves (1895–1985)

ARTHUR KOESTLER

Arthur Koestler (1905–83) was born in Budapest and settled in England in 1940. A prolific journalist, novelist and essayist of strong anti-Fascist persuasion, he was deeply interested in the processes of scientific discovery and artistic creation. His autobiography *Arrow in the Blue* was published in 1952.

from *Arrow in the Blue*

The Pitfalls of Autobiography

Before we go any farther, it may be useful to clarify the question: Why am I writing this autobiography? This should have been done in the preface, but prefaces are so boring to read, and to write, that I have postponed the issue until the story got moving.

I believe that people write autobiographies for two main reasons. The first may be called the 'Chronicler's urge.' The second may be called the '*Ecce Homo*[1] motive.' Both impulses spring from the same source, which is the source of all literature: the desire to share one's experiences with others, and by means of this intimate communication to transcend the isolation of the self.

The Chronicler's urge expresses the need for the sharing of experience related to external events. The *Ecce Homo* motive expresses the same need with regard to internal events.

The Chronicler is driven by the fear that the events of which he is a witness and which are part of his life, their colour, shape, and emotional impact, will be irretrievably lost to the future unless he preserves them on tablets of wax or clay, on parchment or paper, by means of a stylus or quill, typewriter or fountain pen. The Chronicler's urge dominates the autobiographies of persons who themselves have played a part in shaping the history of their times, or felt that they were better equipped than others to record it — as Defoe must have felt when he wrote his *Journal of the Plague Year*.

The *Ecce Homo* motive, on the other hand, urges men to preserve the uniqueness of their inner experiences, and results in the confessional type of autobiography — St. Augustine, Rousseau, de Quincey.[2] It prompts dying physicians to record with minute precision their thoughts and sensations during the last hours before the curtain falls.

Obviously the Chronicler's urge and the *Ecce Homo* motive are at opposite poles on the same scale of values, like introversion and extroversion, perception and contemplation. And obviously a good autobiography ought to be a synthesis of the two — which it rarely is. The vanity of men in public life detracts from the autobiographical value of their chronicles; the introvert's obsession with himself makes him neglect the historical background against which he moves. The *Ecce Homo* motive may degenerate into sterile exhibitionism.

Thus the business of writing autobiography is full of pitfalls. On the one hand, we have the starchy chronicle of the stuffed shirt; on the other, the embarrassing nakedness of the exhibitionist — embarrassing because nakedness is only appealing in a healthy body; who but a doctor wants to look at a rash-covered skin? Apart from these two extremes there are various other snares which even competent craftsmen are rarely able to avoid. The most common of these is what one might call the 'Nostalgic Fallacy.'

With an aching, loving, bitter-sweet nostalgia, the author bends over his past like a woman over the cradle of her child; he whispers to it and rocks it in his arms, blind to the fact that the smiles and howls, and wrigglings of his budding ego lack for his readers that unique fascination which they hold for him. Even experienced authors who know that the reader is a cold fish who has to be tickled behind the gills to make him respond, become victims of this fallacy as soon as they embark on the first chapter with the heading: 'Childhood.' The smell of lavender in mother's linen closet is so intimate; the smile on granny's face so comforting; the water in the brook behind the watercress patch by the garden fence so cool and fresh that it still caresses his fingers holding the pen; and on and on he goes about his linen closets, grannies, ponies, and watercress brooks as if they were a collective memory of all mankind and not, alas, his separate and incommunicable own. Never is the isolation of the self so acutely painful as in the frustrated attempt to share memories of those earliest and most vivid days, when out of the still fluid one-ness of the inside and outside world, out of the original mix-up of fact and fantasy, the sharp boundaries of the self were formed. The Nostalgic Fallacy is the result of the craving to melt and undo those boundaries once again.

The sagacious autobiographer will, therefore, with a sigh of regret, put the dry, crumbling, unique sprig of lavender back in the drawer as if it were a packet of common mothballs and restrict himself to relevant facts. But here the trouble starts again, for how is he to know which facts are relevant and which are not? Both the detective and the psycho-analyst affirm that apparently irrelevant facts yield the most important clues. And my experience with sleuths — whether they searched my pockets or my dreams — has convinced me that by and large the affirmation is correct. When one re-reads the entries in one's diary after five years, one is surprised to find that the most significant events are all strangely under-emphasised. Thus the selection of relevant material is a highly problematical affair, and the crux of all autobiography.

Next among the snares is the 'Dull Dog Fallacy.' A great many memoir-writers are so afraid of showing off that they portray themselves as the dullest dogs on earth. The 'Dull Dog Fallacy' requires that the first person singular in an autobiography should always appear as a shy, restrained, reserved, colourless individual, and the reader wonders how he could possibly succeed in making so many friends, in being always in the midst of interesting people, events, and emotional entanglements. But the Dull Dog is, of course, also a paragon of quiet reliability and unobtrusive decency; if he confesses to certain faults, it is merely an added sign of his modesty.

The virtues of understatement and self-restraint make social intercourse civilised and agreeable, but they have a paralysing effect on autobiography. The memoir-writer ought neither to spare himself nor hide his life under a bushel; he must obviously overcome his reluctance to relate painful and humiliating experiences, but he must also have the less obvious courage to include those experiences which show him in a favourable light.

I do not believe that either in life or in literature puritanism is a virtue. Self-castigation, yes. And self-love too — if it is as fierce and humble, exacting and resigned, accepting and rebellious, and as full of awe and wonder as love for other creatures should be. He who does not love himself, does not love well; and he who does not hate himself, does not hate well; and hatred of evil is as necessary as love if the world is not to come to a standstill. Tolerance is an acquired virtue; indifference is a native vice. 'When I have forgiven a fellow everything, I am through with him,' said Freud. And even Christ hated the moneylenders.

In 1937, during the war in Spain, when I found myself in prison with the prospect of facing a firing squad, I made a vow: if ever I got out of there alive I would write an autobiography so frank and unsparing of myself that it would make Rousseau's *Confessions* and the *Memoirs* of Cellini[3] appear as sheer cant.

That was fifteen years ago; since then I have tried several times to fulfil that vow. I never got farther than the first few pages. The process of self-immolation is certainly painful, but that isn't the real trouble. The trouble is that it is also morbidly pleasant, like the analyst's couch. It leads to the Nostalgic Fallacy in reverse: the scent of the lavender-bag in the drawer is replaced by the sewer smells so dear to our little ids. Moreover, it offers that wrong form of catharsis which the artist learns to avoid like the plague. And whatever is bad art is also bad autobiography. I forced myself to go on because I suspected that my loathing for the job, my revulsion against turning an autobiography into a clinical case-history was due to moral cowardice; and it took me a long time to discover that in this domain the artless truth is obsessional and strident. In short, all art contains a portion of exhibitionism — but exhibitionism is not art.

There is still another aspect to this tricky problem of selecting the relevant material. There is the question: relevant to whom? To the reader, obviously. But what type of reader does the author have in mind? This question, at least, I can answer without ambiguity. The Chronicler's urge is always directed toward the unborn, future reader. This may sound presumptuous, but it is merely the expression of a natural bent. I have no idea whether fifty years from now anybody will want to read a book of mine, but I have a fairly precise idea of what makes me, as a writer, tick. It is the wish to trade a hundred contemporary readers against ten readers in ten years' time and one reader in a hundred years' time. This has always seemed to me what a writer's ambition should be. It is the point where the Chronicler's urge merges with the *Ecce Homo* motive.

Preface to 'High and Low'

MURRAY,[1] you bid my plastic pen
A preface write. Well, here's one then.
Verse seems to me the shortest way
Of saying what one has to say,
A memorable means of dealing
With mood or person, place or feeling.
Anything extra that is given
Is taken as a gift from Heaven.
   The English language has such range,
Such rhymes and half-rhymes, rhythms strange,
And such variety of tone,
It is a music of its own.
With MILTON it has organ power
As loud as bells in Redcliffe[2] tower;
It falls like winter crisp and light
On COWPER's Buckinghamshire night.
It can be gentle as a lake,
Where WORDSWORTH's oars a ripple make
Or rest with TENNYSON at ease
In sibilance of summer seas,
Or languorous as lilies grow,
When DOWSON's lamp is burning low —
For endless changes can be rung
On church-bells of the English tongue.
   MURRAY, your venerable door
Opened to BYRON, CRABBE and MOORE
And TOMMY CAMPBELL. How can I,
A buzzing insubstantial fly,
Compare with them? I do not try,
Pleased simply to be one who shares
An imprint that was also theirs,
And grateful to the people who
Have bought my verses hitherto.

John Betjeman (1906–84)

NORTHROP FRYE

Northrop Frye (1912–91), academic and ecclesiastic, wrote highly influential literary and Biblical criticism. The address *Humanities in a New World*, which discusses issues still very much subject to debate in universities today, was delivered in 1958 to mark the installation of a new President of the University of Toronto, where Frye himself occupied a Chair of English from 1967 until his retirement.

Humanities in a New World

In his satirical romance *Erewhon*, published in 1872, Samuel Butler1 describes the 'Colleges of Unreason,' which taught mainly the 'hypothetical languages,' languages of great difficulty that never existed. The professors were obsessed with the notion that in this world all well-bred people must compromise, hence they instructed their students never to commit themselves on any point. They had professorships of Inconsistency and Evasion, and students were plucked in examinations for a lack of vagueness in their answers. There was however a more modern feeling that examinations should be abolished altogether, the competition involved being regarded as 'self-seeking and unneighbourly.' The strictest of the professors was the professor of Worldly Wisdom, who was also the President of the Society for the Suppression of Useless Knowledge, and for the Completer Obliteration of the Past. Butler concludes that at these Colleges 'The art of sitting gracefully on a fence has never, I should think, been brought to greater perfection.'

The point of Butler's satire is that the more the university tries to remain aloof from society, the more slavishly it will follow the accepted patterns of that society. The tendencies that Butler ridicules are those of a social system in which the ideal is a gentlemanly amateur, with no definite occupation. The university that confronts us today still reflects accepted social attitudes, but those attitudes have changed, and the university has changed with them. The university is now well aware of its social function, and if it were not, public opinion would compel it to become so. Professors are still unwilling to commit themselves, but their reasons are no longer abstract social reasons, but concrete political ones.

Above all, the ideal of productivity, the vision of the unobstructed assembly line, has taken over the university as it has everything else. The professor today is less a learned man than a 'productive scholar.' He is trained in graduate school to become productive by an ingenious but simple device. It is a common academic failing to dream of writing the perfect book, and then, because no achievement can reach perfection, not writing it. One of the major scholarly enterprises on the University of Toronto campus, Professor Kathleen Coburn's edition of the note-books of Coleridge, is the result of the fact that Coleridge never wrote his gigantic masterpiece, the treatise on the Logos that would tell the world what Coleridge knew, but hugged it to his bosom in the form of fifty-seven note-books. *Nous avons changé tout cela.*[2] Our graduate student today must finish a thesis, a document which is, practically by definition, something that nobody particularly wants either to write or to read. This teaches him that it is more important to produce than to perfect, and that it is less anti-social to contribute to knowledge than to possess it.

In Butler's day there were no PhD's in English, but since then there has been a vast increase in the systematizing of scholarship. The modern library, with its stacks and microfilms; modern recordings, reproductions of pictures, aids in learning languages: all these are part of a technological revolution that has transformed the humanities equally with the sciences. There were Canadian poets and novelists a century ago, and critics who reviewed and discussed their work, but there was not the same sense of the systematic processing of the literature that there is now. Of course, wherever there is a cult of productivity there is a good deal of hysteria. New students come along with reputations to make; new poets arise to be commented on; learned journals multiply and their subsidies divide; bibliographies lengthen, and so does the list of works that a scholar feels apologetic about not having read. There seems no answer to this steadily increasing strain on the scholarly economy except the Detroit answer, that next year's books will be still bigger, duller, fuller of superfluous detail, and more difficult to house. If I were speaking only to scholars in the humanities, I should say merely that this is our business, and that we can take care of it. But as I am speaking to a wider public, I should like also to try to explain, if I can, what difference our business makes in the world.

I begin with the fact that the faculty of arts and sciences, or more briefly the faculty of arts, seems to be the centre of the University. A big modern university could almost be defined as whatever group of professional schools in one town happens to be held together by a faculty of arts. We can have a university that is nothing but a faculty of arts; on the other hand, a professional school, set up by itself, is not a university, although it may resemble university life in many ways.

The reason for this is not hard to see as far as science is concerned. The university is the power house of civilization, and the centre of the university has to correspond to the actual centres of human knowledge. Engineering is practical or applied science; medicine is really another form of applied science. And if we ask what it is that gets applied in these professions, the answer is clearly science, as conceived and studied in the faculty of arts. The basis of technology, or applied science, is a disinterested research, carried on without regard to its practical applications, ready to take the risk of being thought useless or socially indifferent or morally neutral, concerned only with developing the science, not with improving the lot of mankind. Technology by itself cannot produce the kind of scientist that it needs for its own development: at any rate, that seems to be the general opinion of those who are qualified to have an opinion on the subject.

Attached to the sciences are what we call the liberal arts or humanities. What are they doing at the centre of university life? Are they there because they must be there, or merely because they have always been there? Are they functional in the modern world, or only ornamental? The simplest way to answer these questions is to go back to the principle on which, in the Middle Ages, the seven liberal arts were divided into two groups. The two great instruments that man has devised for understanding and transforming the world are words and numbers. The humanities are primarily the *verbal* disciplines; the natural sciences are the numerical ones. The natural sciences are concerned very largely with measurement, and at their centre is mathematics, the disinterested study of numbers, or quantitative relationships. At the centre of the humanities, corresponding to mathematics, is language and literature, the disinterested study of words, a study which ranges from phonetics to poetry. Around it, corresponding to the natural sciences, are history and philosophy, which are concerned with the verbal organization of events and ideas.

And just as we have engineering and other forms of applied science, so there is a vast area of what we may call verbal technology, the use of words for practical or useful purposes. The two words practical and useful do not of course mean quite the same thing: some forms of verbal technology, like preaching, may be useful without always being practical; others, like advertising, may be practical without always being useful. Many of the university's professional schools — law, theology, education — are concerned with verbal technology, and so is every area of human knowledge that employs words as well as numbers, metaphors as well as equations, definition as well as measurement. A century ago the central subjects in arts were Classics and mathematics, Classics being restricted to Greek and Latin. Today the central subjects are still Classics and mathematics, but Classics has broadened out to take in all the languages in our cultural orbit, beginning with our own.

This seems clear enough: why are people so confused about the humanities, and more especially confused about literature? There are many answers, but the important one is quite simple. A student who learns only a few pages of Latin grammar will never see the point of having learned even that; and today he learns so little English in early life that the majority of our young people can hardly be said to possess even a native language. 'I think,' said Sir Philip Sidney, '(it) was a piece of the Tower of Babylon's curse, that a man should be put to school to learn his mother tongue.'[3] But it is no use pretending that the curse of Babel does not exist. Behind *Paradise Lost*, behind *Hamlet*, behind *The Faerie Queene*, lay years of daily practice in translating Latin into English, English into Latin, endless themes written and corrected and rewritten, endless copying and imitation of the Classical writers, endless working and reworking of long lists of rhetorical devices with immense Greek names. Discipline of this kind is apparently impossible in the modern school, where teachers are not only overworked but subjected to anti-literary pressures. They are encouraged, sometimes compelled, to substitute various kinds of slick verbal trash for literature; they are bedevilled with audiovisual and other aids to distraction; their curricula are prescribed by a civil service which in its turn responds to pressure from superstitious or prurient voters. In the verbal arts, the student of eighteen is about where he should be at fourteen, apart from what he does on his own with the help of a sympathetic teacher or librarian. To say this is not to reflect on schools, but on the social conditions that cripple them.

So the student often enters college with the notion that reading and writing are elementary subjects that he mastered in childhood. He may never clearly have grasped the fact that there are differences in levels of reading and writing, as there are in mathematics between short division and integral calculus. He is disconcerted to find that, after thirteen years of schooling, he is still, by any civilized standard, illiterate. Further, that a lifetime of study will never bring him to the point at which he has read enough or can write well enough. Still, he is, let us say, an intelligent and interested student with a reasonable amount of good will — most students are, fortunately. He begins to try to write essays, perhaps without ever having written five hundred consecutive words in his life before, and the first results take the form of that verbal muddle which is best called jargon. He is now on the lowest rung of the literary ladder, on a level with the distributors of gobbledygook, double talk and officialese of all kinds; of propaganda, public relations and Timestyle; of the education textbook that is not lucky enough to be rewritten in the publisher's office.

By jargon I do not mean the use of technical terms in a technical subject. Technical language may make one's prose look bristly and forbidding, but if the subject is genuinely specialized there is no way to get out of using it. By jargon I mean writing in which words do not express meanings, but are merely thrown in the general direction of their meanings; writing which can always be cut down by two-thirds without the loss of whatever sense it has. Jargon always unconsciously reveals a personal attitude. There is the blustering jargon that says to the reader, 'Well, anyway, you know what I mean.' Such writing exhibits a kind of squalid arrogance, roughly comparable to placing a spittoon on the opposite side of the room. There is the coy jargon which, like the man with one talent, wants to wrap up and hide away what it says so that no reader will be able to dig it. There is the dithering jargon that is afraid of the period, and jerks along in a series of dashes, a relay race whose torch has long since gone out. There is the morally debased jargon of an easily recognised type of propaganda, with its greasy clotted abstractions, its weaseling arguments, its undertone of menace and abuse. There is the pretentious jargon of those who feel that anything readable must be unscientific. And finally, there is the jargon produced by our poor student, which is often the result simply of a desire to please. If he were studying journalism, he would imitate the jargon of journalism; as he is being asked to write by professors, he produces the kind of verbal cotton-wool which is his idea of the way professors write. What is worse, it is the way that a lot of them do write.

When teachers of the humanities attack and ridicule jargon, they do not do so merely because it offends their aesthetic sensibilities, offensive as it is in that respect. They attack it because they understand the importance of a professional use of words. The natural sciences, we said, are largely concerned with measurement, which means accurate measurement. In any subject that uses words, the words have to be used with precision, clarity and power, otherwise the statements made in them will be either meaningless or untrue. Lawyers, for example, use words in a way different from the poets, but their use of them is precise in their field, as anyone who tries to draft a law without any legal training will soon discover. And what is true of law cannot be less true of sociology or metaphysics or literary criticism.

It is often thought that teachers of the humanities judge everything in words by a pedantic and rather frivolous standard of correctness. They don't care, it is felt, what one really means; all they care about is whether one says 'between you and I,' or uses 'contact' and 'proposition' as verbs. Now it is true that the humanities are based on the accepted forms of grammar, spelling, pronunciation, syntax and meaning. If a man says he will pay you what he owes you next Toisday, it is useful to know whether he means Tuesday or Thursday: if there were no accepted forms there could be no communication. Teachers in the humanities are also concerned with preventing words from being confused with other words, with preserving useful distinctions among words, with trying to make the methods of good writers in the past available for writers today, with trying to steer a civilized course between dictionary dictatorship and mob rule. Some snobbery is bound to be attached to the ability to use words correctly. We hear a good deal about that.

For some reason we hear less about the much greater amount of snob appeal in vulgarity. Most of the people who say 'throwed' instead of 'threw' know well enough that 'threw' is the accepted form, but are not going to be caught talking good grammar. On other social levels there is a feeling that the natural destiny of those who can handle words properly is to form a kind of genteel servant class: ghost writers who turn out books and speeches for the unlettered great; secretaries who translate the gargles and splutters of their bosses into letters written in English; preachers and professors and speakers at clubs who function as middle-class entertainers. Such a conception of society is very like that of a P. G. Wodehouse novel, where the butler speaks in formal nineteenth-century style and his wealthy young master talks like a mentally retarded child. Then again, as Henry James pointed out about fifty years ago in his book *The American Scene*, there was a time when the absorption of the North American male in business led to the domination of all the rest of civilized life by the woman. The result has been that the word culture, which strictly means everything that man has accomplished since he came down out of the trees, has come to acquire a strongly feminine cast. This sense of the word survived in the silly clichés that people use to prevent themselves from thinking, such as 'longhair,' or, most fatuous and slovenly of them all, 'ivory tower,' a phrase which has become popular because it sounds vaguely female and sexual, like a calendar girl in *Esquire*. But this male absorption in business was the product of an expanding economy and weak labour unions: it is now drawing to a close, and in matters of culture the woman is being joined by what Henry James, with his usual delicacy, called her sleeping partner.

Our student, with a little practice, will soon advance from jargon to the beginnings of prose, which means advancing from an amateurish to a professional approach to words. To make such an advance involves an important moral and psychological change. Bad writers are like bad car-drivers: what they are doing is the unconscious expression of a way of life. The purveyors of jargon are like the man who honks and hustles his way through traffic to advertise the importance of his business, or the woman who wants to hit something in order to prove that she is helpless and appealing. The good car-driver regards his or her activity as a simple but highly specific skill, unconnected with the rest of the personality. The good writer is the writer who puts self-expression aside, and is ready to submit himself to the discipline of words.

In the past, and under the influence of the old faculty psychology, the different fields of study were correlated with different parts of the mind. Thus history was ascribed to the memory, poetry to the imagination, and philosophy and science to the reason. This way of thinking has left many traces in our day: it is still widely believed that a mathematician is an unemotional reasoner, and a poet a 'genius,' a word which usually means emotionally unbalanced. But, of course, any difficult study demands the whole mind, not pieces of it. Reason and a sense of fact are as important to the novelist as they are to the chemist; genius and creative imagination play the same role in mathematics that they do in poetry. A similar fallacy may be confusing our student at this critical point. I am, he perhaps feels, a conscious being; I know I can think; I know I have ideas that are waiting to be put into words. I wish somebody would show me how to express my ideas, instead of shoving all this poetry stuff at me. After all, poets put their *feelings* into words, so they can make sounds and pictures out of them; but that isn't what I want.

Every step in this chain of reasoning is wrong, so it is no wonder if the reasoner is confused. In the first place, thinking is not a natural process like eating or sleeping. The difficulty here is partly semantic: we are apt to speak of all our mental processes as forms of thought. Musing, day-dreaming, remembering, worrying: every slop and gurgle of our mental sewers we call thinking. If we are asked a question and can only guess at the answer, we begin with the words 'I think.' But real thinking is an acquired skill founded on practice and habit, like playing the piano, and how well we can think at any given time will depend on how much of it we have already done. Nor can we think at random: we can only add one more idea to the body of something we have already thought about. In fact, we, as individuals or egos, can hardly be said to think at all: we link our minds to an objective body of thought, follow its facts and processes, and finally, if the links are strong enough, our minds become a place where something new in the body of thought comes to light.

It is the same with the imaginative thinking of literature. The great writer seldom regards himself as a personality with something to say: his mind to him is simply a place where something happens to words. T. S. Eliot compares the poet to a catalyzer, which accompanies but does not bring about the process it is used for; Keats speaks of the poet's negative capability; Wordsworth of his recollection in tranquillity; Milton of the dictation of unpremeditated verse by a Muse. The place where the greatest fusions of words have occurred in English was in the mind of Shakespeare, and Shakespeare, as a personality, was so self-effacing that he has irritated some people into a frenzy of trying to prove that he never existed.

If the student were studying natural science, he would grasp this principle of objective thought very quickly. There can be no self-expressive approach to physics or chemistry: one has to learn the laws of the science first before one can have anything to express in it. But the same thing is true of the verbal disciplines. The student is not really struggling with his own ideas, but with the laws and principles of words. In any process of genuine thought that involves words, there can be no such thing as an inarticulate idea waiting to have words put around it. The words are the forms of which the ideas are the content, and until the words have been found, the idea does not exist.

A student of engineering may have extremely practical aims in entering that field, but he cannot get far without mathematics. Hence mathematics, though not in itself a practical subject, is practical enough for him. For a student who is going to engage in any verbal activity, the study of literature, not in itself a practical subject, is a practical necessity. The sciences deal with facts and truths, but mathematics sets one free from the particular case: it leads us from three apples to three, and from a square field to a square. Literature has the same function in the humanities. The historian is concerned with finding the right words for the facts; the philosopher, with finding the right words for the truth. As compared with the historian, the poet is concerned, Aristotle tells us, not with what happened but with the kind of thing that does happen. As compared with the philosopher, the poet is concerned, not with specific statements, but with images, metaphors, symbols and verbal patterns out of which all directed thinking comes. Mathematics is useful, but pure mathematics, apart from its use, is one of the major disciplines of beauty. Poetry, is, in itself, beautiful, but if we think of it as merely decorative or emotional, that is because we have not learned to use it. We can build the most gigantic structures out of words and numbers, but we have constantly to return to literature and mathematics, because they show us the infinite possibilities that there are in words and numbers themselves. Sir James Jeans, speaking of the failure of nineteenth-century physics to build a mechanical model of the universe, says that the Supreme Architect of the universe must be a mathematician. A much older authority informs us that the Supreme Teacher of mankind was a story-teller, who never taught without a parable.

The humanities in the university are supposed to be concerned with criticism and scholarship, not with creation as such. At the centre of literature lie the 'classics,' the works that university teachers know they can respect, and the university student, *qua* student, is there to study them, not to write on his own. True, most writers of importance today are not only university graduates but university employees, at least in summer sessions. True, the untaught writer who sends a masterpiece to a publisher from out of nowhere is much more a figure of folklore than of actual literature. Still, the university does not try to foster the social conditions under which great literature can be produced. In the first place, we do not know what these conditions are; in the second place, we have no reason to suppose that they are good conditions. Just as doctors are busiest in an epidemic, so our dramatists and novelists may find their best subjects where decadence, brutality, or idiocy show human behaviour in its more fundamental patterns. Or the producer of literature himself may be a drunk, a homosexual, a Fascist, a philanderer; in short, he may want things that the university cannot guarantee to supply.

The university, therefore, addresses itself to the consumer of literature, not to the producer. The consumer of literature is the cultivated man, the man of liberal education and disciplined taste, for whose benefit the poet has worked, suffered, despaired, or even wrecked his life. What the university does try to do is foster the social conditions under which literature can be appreciated. Many teachers of the humanities are anxious to stop at that point, especially those who wish that they have been great poets instead. It is natural for them to insist that critics and scholars have no real function except to brush off the poet's hat and hand it to him. But a merely passive appreciation of literature is not enough. As Gerard Manley Hopkins said: 'The effect of studying masterpieces is to make me admire and do otherwise.'[4] He was a poet, but he has exactly defined, even for non-poets, the effect of great writing, which is great because it is infinitely suggestive, and encourages us not to imitate it, but to do what we can in our own way. To appreciate literature is also to use it, to absorb it into our own lives and activities. There is unlikely to be much of a gap between what the humanities will do in a new world and what they are trying to do in this one. Teachers of the humanities understand the importance of what they are doing, and in any new world worth living in, nine-tenths of their effort would be to go on doing it. Still, I think they will become increasingly interested in the ways in which words and verbal patterns do affect human lives. They are likely to follow the direction indicated by the poet Wallace Stevens in one of his long discursive poems:

> This endlessly elaborating poem
> Displays the theory of poetry
> As the life of poetry. A more severe,
>
> More harassing master would extemporize
> Subtler, more urgent proof that the theory
> Of poetry is the theory of life.[5]

A few years ago there was a great vogue for something called 'semantics,' which purported to be, not simply a certain type of literary study, but a panacea for human ills. People get neurotic, we were told, by attaching private and emotional significances to words: once they learn to use words properly, to bring them into alignment with the world around them, their psychological distresses and tensions will clear up. A minor advantage would be the abolishing of literature, where words are so thickly coated with emotional associations. Like other miraculous cures, semantics of this type achieved a great success among the hysterical, but failed to do everything it promised to do. It looks as though, as long as men are discussing matters that affect their pocketbooks, their homes or their lives, they will continue to attach emotional significance to the words they use. Perhaps it would be better to recognize that there is no short cut to verbal accuracy, and go back to study the poets, who have not tried to get rid of emotion, but have tried to make it precise. Nevertheless, the semanticists were right about the importance of words in human life, about the immediacy and intimacy of their impact, about their vast powers for good or for evil.

We use words in two ways: to make statements and arguments and convey information, or what passes as such, and to appeal to the imagination. The former is the province of history, philosophy and the social sciences; the latter is the province of literature. There is also a large intermediate area of what is called rhetoric, the art of verbal persuasion, where both means are employed. We are brought up to believe that words stand for things, and that most of our experience with words takes the form of reported fact, argument, and logical inference. This is a flattering self-delusion. Most of our daily experience with words takes place on a low level of the imagination — that is, it is sub-literary. I am writing this on the subway, and my eye falls on an advertisement for heavy-duty floor wax. Nothing could be more honestly factual; but even here 'heavy duty' is a metaphor, probably of military origin, and the metaphor, with its imaginative overtones of ruggedness, strength and endurance, is the focus of the sales appeal. If the advertiser has something expensive to sell, this sub-literary appeal is stepped up. One cannot read far in advertising without encountering over-writing, a too earnestly didactic tone, an uncritical acceptance of snobbish standards, and obtrusive sexual symbolism. These are precisely the qualities of inferior literature.

Then there are the other sub-literary areas of soap-operas, movies, magazine stories, jokes, comic strips, gossip. It is out of the steady rain of imaginative impressions from these and similar quarters that most people form their myths: that is, their notions of representative human situations, of typical human characters and characteristics, of what is inspiring and what is ridiculous, of the socially acceptable and the socially outcast. It is here that the kind of preferences develop which determine one to condemn or condone segregation, to support or decry the United Nations, to vote for Mr. Diefenbaker or for Mr. Pearson. For even election issues and current events reach us chiefly through human-interest stories and personal impressions. For better or worse, it is through his literary imagination, such as it is, that modern man participates in society.

The responsible citizen of course, tries to get away from mythical stereotypes, to read better papers and seek out friends who have some respect for facts and for rational discussion. But he will never succeed in raising his standards unless he educates his imagination too, for nothing can drive bad literature out of the mind except good literature. In these days we have an exaggerated sense of the power of argument and indoctrination. 'Ideas are weapons' was a once fashionable phrase, and during the war publishers carried the slogan 'books are weapons in the war of ideas.' But arguments and aggressive ideas have a very limited role to play in human life. They that take the argument will perish by the argument; any statement that can be argued about at all can be refuted. The natural response to indoctrination is resistance, and nothing will make it successful except a well organized secret police. What can never be refuted is the underlying vision of life which all such arguments try to rationalize. The arguments are based on assumptions about what is worth living for or dying for; these are rooted in the imagination, and only the imagination can nourish them.

The distinction that we have made between the disciplines of words and numbers does not quite correspond to the distinction between the arts and the sciences. There are arts that do not depend on words, like music and painting, and there are sciences that do, like the social sciences. The real difference between art and science is expressed by Francis Bacon in *The Advancement of Learning*:

> The use of (poetry) hath been to give some shadow of satisfaction to the mind of Man in those points where the Nature of things doth deny it, the world being in proportion inferior to the soul . . . And therefore (poetry) was ever thought to have some participation of divineness, because it doth raise and erect the Mind, by submitting the shews of things to the desires of the Mind, whereas reason doth buckle and bow the Mind unto the Nature of things.[6]

The sciences, in other words, are primarily concerned with the world as it is: the arts are primarily concerned with the world that man wants to live in. The sciences have among other things the function of showing man how much he can realize of what he wants to do, and how much has to remain on the level of wish or fantasy. In between comes the area of applied science and applied art, where the process of realization is accomplished. Architecture is one obvious place in which science and art meet on a practical basis. Art, then, owes its existence to man's dissatisfaction with nature and his desire to transform the physical world into a human one. Religion itself, when it deals with ultimate things, uses the language of art, and speaks of an eternal city and a restored garden as the fulfilling of the soul's desires. The human imagination, which the arts address, is not an escape from reality, but a vision of the world in its human form.

Science continually evolves and improves: the scientist contributes to an expanding body of knowledge, and the freshman studying physics today can sit on the shoulders of Newton and Faraday, knowing things that they did not know. The arts, on the other hand, produce the classic or model, which may be equalled by something different, but is never improved on. The greatest artists have reached the limits of what their art can do: there is an infinite number of limits to be reached, and artists of the future will reach many of them, but it makes nonsense of the conception of art to think of it as developing. The painters in the stone age caverns were as highly developed as Picasso; Homer is as much a model for poets today as he was for Virgil. We have as great art as humanity can ever produce with us now. The natural direction of science, then, is onward: it moves toward still greater achievements in the future. The arts have this in common with religion, that their direction is not onward into the future but upward from where we stand.

The point of contact between the arts and the human mind is the moment of leisure, one of the most misunderstood words in the language. Leisure is not idleness, which is neurotic, and still less is it distraction, which is psychotic. Leisure begins in that moment of consciousness peculiar to a rational being, when we become aware of our own existence and can watch ourselves act, when we have time to think of the worth and purpose of what we are doing, to compare it with what we might or would rather be doing. It is the moment of the birth of human freedom, when we are able to subject what is actual to the standard of what is possible. William Blake calls it the moment in the day that Satan cannot find. It is a terrifying moment for many of us, like the opening of a Last Judgement in the soul, and our highways and television sets are crowded with people who are not seeking leisure but are running away from it. The same is true of the compulsive worker, the man who boasts of how little leisure he has, and who speeds himself up until he explodes in neuroses and stomach ulcers.

We tend to think of leisure as having nothing in particular to do: this is what the word means in Thorstein Veblen's *Theory of the Leisure Class*, where he is examining the traditional idea of the gentleman as the man who does not work. But even the old leisure class did possess some essential social values — courtesy, good taste, patronage of the arts — and a democracy has the problem of trying to retain those genuine values, while making them accessible to anyone. It is still true that liberal education is the education of the free man, and has no meaning out of the context of freedom. The really privileged person is not the man who has no work to do, but the man who works freely, and has voluntarily assumed his duties in the light of his conception of himself and his social function. The underprivileged person is (at best) the servile worker, or what Carlyle called the drudge, and every social advance, every technological invention, every improvement in labour relations, has the aim of reducing the amount of servile work in society. But what makes free work free is its relation to the vision of life that begins in the moment of leisure. The poet Yeats took as a motto for one of his books the phrase 'in dreams begin responsibilities.' [7] It is also in man's dream of a humanized world that all learning, art and science begin. As Aristotle pointed out, the words school and scholarship come from *schole*, leisure. The Bible says that leisure is the beginning of wisdom: it also says that the fear of the Lord is the beginning of wisdom but the two statements are quite compatible, for religion too has its origin in leisure. Christianity illustrated this fact when it changed a day of rest at the end of the week into a day of leisure at the beginning of the week. The university illustrates the same principle, in its secular form, when it places a four-year voluntary liberal education at the beginning of adult life.

We also tend to think of the rewards of leisure as individual possessions, like the love of poetry or music that fills the spare intervals of our lives with private moments of grace and beauty. But behind these private possessions lies a social possession, a vision of life that we share with others. This shared vision is the total form of art, man's vision of a human world, to which every individual work of art belongs. Most of us are seldom aware of the power of words in forming the visions which hold society together. Special occasions, like the familiar words spoken at marriages and funerals, or a critical moment in history that we happen to live through, like the summer of 1940 when the free world had practically nothing but Churchill's prose style left to fight with, are usually all that bring them to our minds.

Yet any newspaper can show us how society turns on the hinges of words and numbers. The people who make fortunes out of uranium stocks owe their wealth and social prestige to an absent-minded professor, badly in need of a haircut, who scribbled down $e=mc^2$ on a piece of paper fifty years ago. The Republicans owe their existence to the fact that a century ago a long-legged Illinois lawyer put a few words together that made up a social vision for the American people of genuine dignity and power, and so enabled the Republican party to stand for something. Communism owes its existence to the fact that a century ago a carbuncular political agitator disappeared into the British Museum to write a sprawling, badly organized and grittily technical book on capitalism, which even its author was unable to finish. I have no doubt that the philosophy and economics of that work have been refuted many times, but no refutation will have any effect on it. Marxism is a vision of life, with its roots in the social imagination, and it will endure, at least as a vision, until another of greater intensity grows up in its place.

The people who run away from their own leisure will, of course, also run away from the articulate sounds of words that would recall them to their dreams and their responsibilities. Just as a frightened child may be reassured to hear the murmur of his parents' voices downstairs, so the childish in our society turn to the books and newspapers, the television programmes and the political leaders, that supply them with the endless, unmeaning babble of the lonely crowd. If you remember George Orwell's *1984*, you will recall the decisive role of 'Newspeak' in that book. There is only one way to degrade mankind permanently, and that is to destroy language. The whole history of ordered public speech, from the Hebrew prophets who denounced their kings and the Demosthenes and Cicero who fought for the Classical republics down through Milton and Jefferson and Mill and Lincoln, has been inseparably a part of the heritage of freedom. In the nature of things — or rather in the nature of words — it cannot be otherwise. We naturally demand leadership from our leaders, but thugs and gangsters can give us leadership, of a kind: if we demand articulateness as well, we are demanding something that only a genuine vision of human life can provide.

Word

The word bites like a fish.
Shall I throw it back free
Arrowing to the sea
Where thoughts lash tail and fin?
Or shall I pull it in
To rhyme upon a dish?

Stephen Spender (b. 1909)

DORIS LESSING

In the preface to *Shikasta*, the first of her 'science fiction' novels, Doris Lessing describes the freedom she discovered in moving away from the more traditional novel-form she had hitherto followed. Her closing observations are worth comparing with Robert Graves's poem '*To Juan at the Winter Solstice*' (p. 191).

Some Remarks (from *Preface to 'Shikasta'*)

S hikasta was started in the belief that it would be a single self-contained book, and that when it was finished I would be done with the subject. But as I wrote I was invaded with ideas for other books, other stories, and the exhilaration that comes from being set free into a larger scope, with more capacious possibilities and themes. It was clear I had made — or found — a new world for myself, a realm where the petty fates of planets, let alone individuals, are only aspects of cosmic evolution expressed in the rivalries and interactions of great galactic Empires: Canopus, Sirius, and their enemy, the Empire Puttiora, with its criminal planet Shammat. I feel as if I have been set free both to be as experimental as I like, and as traditional: the next volume in this series, *The Marriages Between Zones Three, Four, and Five*, has turned out to be a fable, or myth. Also, oddly enough, to be more realistic.

It is by now commonplace to say that novelists everywhere are breaking the bonds of the realistic novel because what we all see around us becomes daily wilder, more fantastic, incredible. Once, and not so long ago, novelists might have been accused of exaggerating, or dealing overmuch in coincidence or the improbable: now novelists themselves can be heard complaining that fact can be counted on to match our wildest inventions.

As an example, in *The Memoirs of a Survivor* I 'invented' an animal that was half-cat and half-dog, and then read that scientists were experimenting on this hybrid.

Yes, I do believe that it is possible, and not only for novelists, to 'plug in' to an overmind, or Ur-mind, or unconscious, or what you will, and that this accounts for a great many improbabilities and 'coincidences.'

The old 'realistic' novel is being changed, too, because of influences from that genre loosely described as space fiction. Some people regret this. I was in the States, giving a talk, and the professor who was acting as chairwoman, and whose only fault was that perhaps she had fed too long on the pieties of academia, interrupted me with: 'If I had you in my class you'd never get away with that!' (Of course it is not everyone who finds this funny.) I had been saying that space fiction, with science fiction, makes up the most original branch of literature now; it is inventive and witty; it has already enlivened all kinds of writing; and that literary academics and pundits are much to blame for patronising or ignoring it — while of course by their nature they can be expected to do no other. This view shows signs of becoming the stuff of orthodoxy.

I do think there is something very wrong with an attitude that puts a 'serious' novel on one shelf and, let's say, *First and Last Men*[1] on another.

What a phenomenon it has been — science fiction, space fiction — exploding out of nowhere, unexpectedly of course, as always happens when the human mind is being forced to expand: this time starwards, galaxy-wise, and who knows where next. These dazzlers have mapped our world, or worlds, for us, have told us what is going on and in ways no one else has done, have described our nasty present long ago, when it was still the future and the official scientific spokesmen were saying that all manner of things now happening were impossible — who have played the indispensible and (at least at the start) thankless role of the despised illegitimate son who can afford to tell truths the respectable siblings either do not dare, or more likely, do not notice because of their respectability. They have also explored the sacred literatures of the world in the same bold way they take scientific and social possibilities to their logical conclusions so that we may examine them. How very much we do all owe them!

Shikasta has as its starting point, like many others of the genre, the Old Testament. It is our habit to dismiss the Old Testament altogether because Jehovah, or Jahve, does not think or behave like a social worker. H. G. Wells said that when man cries out his little 'gimme, gimme, gimme' to God, it is as if a leveret were to snuggle up to a lion on a dark night. Or something to that effect.

The sacred literatures of all races and nations have many things in common. Almost as if they can be regarded as the products of a single mind. It is possible we make a mistake when we dismiss them as quaint fossils from a dead past.

Leaving aside the Popol Vuh, or the religious traditions of the Dogon, or the story of Gilgamesh,[2] or any others of the now plentifully and easily available records (I sometimes wonder if the young realise how extraordinary a time this is, and one that may not last, when any book one may think of is there to be bought on a near shelf) and sticking to our local tradition and heritage, it is an exercise not without interest to read the Old Testament — which of course includes the Torah of the Jews — and the Apocrypha, together with any other works of the kind you may come on which have at various times and places been cursed or banished or pronounced non-books; and after that the New Testament, and then the Koran. There are even those who have come to believe that there has never been more than one Book in the Middle East.

August, 1968 [1]

The Ogre does what ogres can,
Deeds quite impossible for Man,
But one prize is beyond his reach,
The Ogre cannot master Speech:
About a subjugated plain,
Among its desperate and slain,
The Ogre stalks with hands on hips,
While drivel gushes from his lips.

W. H. Auden (1907–73)

WITI IHIMAERA

Witi Ihimaera, Maori novelist and short-story writer, analyses his role (and by implication that of writers in general) in a piece whose title echoes George Orwell's essay printed earlier in this collection.

Why I Write

There are two cultural maps of my country, the Maori and the Pakeha. The Pakeha map is dominant, its contours so firmly established that all New Zealanders including Maori are shaped by it. The Maori map has eroded and, although its emotional landscape is still to all intents and purposes intact, has been unable to shape all New Zealanders including Pakeha.

Although the situation is improving, the erosion lessening, most New Zealanders remain unaware that they have a dual cultural heritage and not a single one. Their attitude is still predominantly separatist, which is surprising for a country which prides itself for its amicable record of race relations. For instance, ask who discovered New Zealand and you will be told Abel Tasman. But the answer, as given by Maori history, is Kupe.

And that, quite simply, is why I began to write. To make New Zealanders aware of their 'other,' Maori, heritage. To convince my countrymen, with love and anger, that they must take their Maori personality into account.

My way is with the pen; others of my people use more forceful methods. Mine is not necessarily an activist approach and I have been accused of not being 'political' enough or critical enough of our Pakeha-dominated society or hitting hard enough at the very real social, economic, legal and other problems facing the Maori people today. Okay. But I say my work is political because it is exclusively Maori, the criticism of Pakeha society is implicit in the presentation of an exclusively Maori values system, and I am more concerned with the greatest problem we have — that of retaining our emotional identity — rather than the more individual but also very real social problems some of us face today. My concern is for the roots of our culture, the culture we carry within ourselves and which makes us truly Maori. It is a culture essentially rurally based, with toes firmly gripping the soil, and so I write about the rural Maori rather than about the Maori in urban areas where — so I've been informed — all the action is. But only when I have completed writing about the rural Maori to my satisfaction will I uncurl my toes and write about how hard the city pavements are to our feet. Not before. Not yet.

I write about the landscapes of the heart, the emotional landscapes which make Maori people what they are. My first priority is to the young Maori, the ones who have suffered most with the erosion of the Maori map, the ones who are Maori by colour but who have no emotional identity as Maori. My second priority is the Pakeha — he must understand his Maori heritage, must understand that cultural difference is not a bad thing and that, in spite of the difference, he can incorporate the Maori vision of life into his own personality. Thirdly, I write for all New Zealanders to make them aware of the tremendous value in Maori culture and of the tragedy for them should they continue to disregard this part of their dual heritage. They can accept or reject that heritage if they want to, but they must at least have the opportunity to choose. If some choose to reject it, fair enough, I won't cry at their tangi.[1] But allow Maori people at least the right to maintain their heritage and their individuality.

Conveying this heritage through writing has not been easy. I have had to answer to both Maori and Pakeha for what I am doing and have hurt both, sometimes unintentionally but most often with the intention of waking them up; our own tradition of oral literature is not enough to carry our culture into the future, and a certain amount of force has had to be applied to my Maori elders on one hand and Pakeha editors on the other to allow opportunities for the Maori written word, even if the results do make them wince. As well, there has been some pressure from people who want me to write what they want; I hope that writers who are Maori never succumb to any expectations other than the ones they carry themselves.

But the main problem is that the writer who is Maori has both a dual role as a writer and a Maori and a dual responsibility to his craft and his people. Even if he does not see himself this way, you can bet your life that other people do and indeed are very eager to cast him in various roles. If he is strong-willed enough, he must ensure that no patronisation is given his art as being good . . . for a Maori, that is . . . for he is a writer first and a Maori second. However, he is also a Maori writer and must suffer patronisation because people are often more interested in him as a Maori and not as a writer. It is difficult to balance both roles and both responsibilities. Certainly, I have found it so, but there seems no way out. I am committed both to my people and my art as a writer.

My main fear? Mine is only an individual response to Maori life and should not be taken as the definitive view. It can only be an interim report, is in danger of setting up a stereotype of its own, and I look forward to the emergence of more writers who are Maori. Only then can the broad spectrum of Maori experience become available and the Maori map become fully drawn. Not until then will New Zealanders become fully aware of their dual heritage of Maori and Pakeha culture.

Tena koutou, tena koutou, tena koutou katoa.

The Thought-Fox

I imagine this midnight moment's forest:
Something else is alive
Beside the clock's loneliness
And this blank page where my fingers move.

Through the window I see no star:
Something more near
Though deeper within darkness
Is entering the loneliness:

Cold, delicately as the dark snow,
A fox's nose touches twig, leaf;
Two eyes serve a movement, that now
And again now, and now, and now

Sets neat prints into the snow
Between trees, and warily a lame
Shadow lags by stump and in hollow
Of a body that is bold to come

Across clearings, an eye,
A widening deepening greenness,
Brilliantly, concentratedly,
Coming about its own business

Till, with a sudden sharp hot stink of fox
It enters the dark hole of the head.
The window is starless still; the clock ticks,
The page is printed.

Ted Hughes (b. 1930)

JANET FRAME

Since writing this brief memoir Janet Frame has described her early life and writing career far more fully in the three volumes of her autobiography — *To the Is-land, An Angel at My Table* and *The Envoy from Mirror City*. The 'condensed' account here nevertheless raises issues of considerable importance and interest about writing.

Beginnings

A re there 'pockets' of poetry in the world as there are 'pockets' of depression and wealth, areas breeding poetry like a rare plant which the nation eats to satisfy an extra appetite, enjoying the pleasant taste without thinking too much of the dangers of the 'insane root'?

I was born into a family, local and national 'pocket' of poetry. In my family words were revered as instruments of magic. My father's mother whose cross on her marriage certificate had been witnessed as 'Mary Paterson, her mark' used all the forces of her Scottish superstition to try to cope with the sudden family literacy, the eleven children who were born and going to school in the new land, and learning to read and write words. In a recent visit to my old home to sort family papers and letters I've seen the legacy of my grandmother's superstition in the way my father carefully hoarded grocery lists, greeting cards, notes, receipts, certificates, licences, telegrams, pages of financial plotting rich with £££ signs, Art Union and Tatts tickets: the debris of a revolution of literacy, millions of words whose power to destroy, to revive, transform, had been so respected and feared that the words had been kept prisoners, almost as if my family expected the telegrams of many years past to present themselves again at the front door, provoking renewed shock — joy, grief. As my mother's family had been literate for many generations and she had not suffered this devastating avalanche of words, her methods of controlling them were less disorganized. Even her great grandmother had found a solution to the problem of the power of words by shaping them as poems into notebooks. Writing poems became a family habit, though some of the succeeding generations chose to make little books of household remedies rather than poems. My mother united the two: she wrote poems as remedies for her own ills and those of her family, her friends and her country. A dreamer, who never gave up her romantic notion that great writers compose masterpieces on backs of envelopes, receipts, old letters, she left each day a trail of poem-spattered fragments in the hope, perhaps, that the gods would favour her for meeting the first condition of literary greatness. So my father sat at one end of the table writing his time-sheets, recording pack and trim, draining tanks and tenders, tablet exchanges; my mother sat at the other end writing her poems to submit for publication, while my sisters and I sat on the long kauri form or the bin near the fire writing *our* poems to send to the newspapers. We were proud to have our work printed, and read over the air. When I began to share my interest in poetry with a Canadian penfriend who sent me half a dollar, a maple-leaf which became brittle and crumbled, and a book of verse by the Young Co-operators Club, full of evocative references to prairies, trappers, snow; spruce, tamrac, quaking muskeg beds — then my chosen world — North America — was complete. Much has been written of the English background to the life of a New Zealander; little has been said of the North American influence. Though my mother and father had been born in New Zealand (indeed, my step-great-grandmother was a full-blooded Maori) and drew their themes for bedtime stories from life around the Port Chalmers Wharf and 'down the Sounds' in Marlborough, my Scottish grandmother convinced me, by her tales of life in America, and by her knowledge and singing of negro spirituals, that my ancestral home was North America. I do not know where my grandmother learned her tales and songs; I believed that she had come from America, that she had been a slave there, for she was big, with frizzy hair and dark skin, and sometimes when she sang about the cotton fields she would cry as if she were remembering her years as a slave.

As a child then, I dreamed more often of prairies than of paddocks, of coyotes than wallabies, and in winter when the bitter wind blew from the Tokorahis (or were they the Rockies?) I had to struggle through deep snowdrifts down town to buy — *The Otago Daily Times. . . .*

I made my first story on the banks of the Mataura river after a meal of trout and billy tea. 'Once upon a time there was a bird. One day a hawk came out of the sky and ate the bird. The next day a big bogie came out from behind the hill and ate up the hawk for eating up the bird.'

The story's not unusual, told by a child of three. As I still write stories I'm entitled to study this and judge it the best I've ever written. When my Scottish grandmother came to one of the most memorable moments of her life she had *no words* to waste in writing. When I was small, and the hawks so often came swooping out of the sky to kill the birds and the bogies so often lived behind the hills, and I thought I'd make up a story to put hawk, bird, bogie into their proper hierarchy, then I had few words to waste. Getting them, using them were as simple as being the hawk which swooped out of the sky. I keep that story in mind as an example of a time in my life when I did not waste words, when I had fewer words to choose from, yet I was still unable to perform the miracle of keeping my listeners 'from play' and (as equivalent of the 'chimney-corner')[1] from the paddling pool — clear water on a bed of white stones — near the river. I remember that I called on the authority of my mother — Mum, they're wriggling, stop them from wriggling while I tell a story. How much more difficult it is now, to choose words that bind with their spell, and to realize that once the words are chosen and written, there's no one to appeal to, to help recall the wandering attention! One is quite alone, one must decide alone. Recently, for the first time in seven years, I travelled in a New Zealand train which stopped suddenly in a country wilderness, not, it seemed, for the accepted reasons — to get water, to let the north express or a goods train pass; there was no reason that I could think of, and the code forbade me to ask, for I was of the country, wasn't I, and should have known? When I write, and my words go astray, with unexplained stops and 'wrigglings' of concentration in those who read, I think to myself — I'm of the country — the country of words — I should have known.

The Second War came. Books were printed on thin yellow paper with smudges and specks on it; such thin paper that it seemed unable to bear the weight or shape of the words; and where could words be contained if paper could no longer house them? I felt as if the floor and walls of the world were being torn apart; there seemed to be nowhere anymore, no hope, once governments had decreed the rationing of paper and books. The books in my life had once been few. On the shelves in the kitchen at home there were none of the books which my mother was able to remember and recite from: Hawthorne, Mark Twain, Dickens, and the American poets. My mother had come from a home where there were many books to one where there were few; in a way, her marriage was a migration; she retained passages of prose and poetry and recited them as if they were vivid memories of a homeland she would never see again. Our bookshelf had Grimm's *Fairy Tales* with its dark small print enhancing the terror of many of the tales, and with occasional pages stiffened and curled as if they had been exposed to the weather, as they had been, for Grimm's *Fairy Tales* and Ernest Dowson's *Poems*, and George Macdonald's *At the Back of the North Wind* had been found in the town rubbish dump. The other family books were twelve volumes of Oscar Wilde (which my father bought at an auction in Wyndham, in a 'lot' which included Oscar Wilde, a yellow and black chiming clock, a pair of hedge shears, and a bagpipe record, The Wee Macgregor), *Christendom Astray*, my mother's Christadelphian manual, the Bible, God's Book (a luridly illustrated account of the creation and the prophets and the Latter Days), the doctor's Book, an equally luridly illustrated account of the human body in sickness; a collection of school books and prizes, and one 'foreigner' which I never read: it was pale blue with white stripes down the cover and spine, and its title was *To Pay the Price*.

When I was Dux of the primary school I was given a medal and a library subscription for the following year. The whole family shared in the wonders of the books I brought home, and when the subscription had finished we agreed that no matter how hard it was to make ends meet at home, we could not give up the pleasures of reading. My mother (excited at the thought of communicating with the characters and poems of her past) begged, 'Bring home Charles Dickens, bring home Nathaniel Hawthorne, Henry Wadsworth Longfellow, John Greenleaf Whittier, Mark Twain (that is, kiddies, Samuel Clemens).' My mother would never dream of giving a writer anything but his or her full Christian and surname. I dutifully 'brought home' the favourites including 'a William book' for my brother who was ill, and 'something about the sea' for my father. The library tradition was begun. As my ambition from the time I was eight was to be a poet, and I was now at High School, in preparation for my future profession I read chiefly poetry and books about poetry and poets. Along a lonely wall of the town library (I felt it belonged to me, for no one seemed to choose books from there) I found the Complete Works of the Poets, all the Victorian Novelists, German Philosophers (I could never be a poet without having studied Kant), the translated works of Georges Sand (I descended from my poetic pedestal to cherish a secret passion for Count Albert in *Consuelo*). I was almost in despair because I could never find Chapman's Homer to 'look into', nor Holinshed's histories to read the original stories of *Macbeth* and *King Lear*.

In our later years at High School my sisters and I, becoming more ambitious, set to work on our novels, with the titles carefully chosen: *There is Sweet Music*, *Go Shepherd*, and *The Vision of the Dust*, which I had chosen because I was currently anti-shepherd, anti-sweet music, seeking the poetry in 'the heart of the unobvious' by writing about such topics as cellophane paper, factories, ditches, slums, ugliness. There were tragic happenings in our family. Sometimes when it seemed to us that our family was doomed, we would console ourselves by remembering the Brontës and drawing between them and us a grandiose dramatic parallel which could not harm, though it might have amused, them, but which may have harmed us into believing that we three girls held, by right, 'silk purses' of words. With a background of poverty, drunkenness, attempted murder and near-madness, it was inevitable that we should feel close to the Brontës, once we had read their books and knew the story of their lives. My younger sister (who later died when she was twenty) was assigned the role of Emily; I, more practical and less outwardly 'passionate' became Charlotte, while my youngest sister, shy, over-shadowed in many ways by our 'glory', became Anne.

Then, when my own private world became more demanding and I no longer cared to confide in 'Emily' or 'Anne' I spent my spare time writing my diary addressed to the ruler of my Land of Ardenue. I lived increasingly in this imaginary world whose characters were drawn from objects and people I met in my daily life, with occasional intrusion of characters from fiction — from Dostoyevsky, Daudet, Hardy. My home task of milking the two cows enabled me to spend hours on the hills around Oamaru talking and exchanging opinions with my characters while I persuaded the cows (which

I milked in the paddocks) to 'let down' their milk. Or I made poems in my head and wrote them when I came home.

The channels of poetry were strengthened and deepened. Those needs and longings which, during childhood, had been stacked in a tremendous slow-moving mass, like a picturesque glacier seen in the distant mountains of the high country, had begun to thaw, in the spring of adolescence. One always remembers vividly those chilling crashing immense movements of the spirit as the long-frozen ideas found their life and course and merged one with the other in their flow to the sea.

People began asking, What are you going to do with your life? I'll be a teacher, I said, writing in my diary the same evening, I'm going to be a poet.

I left home to attend Training College and University in Dunedin. I looked for the poets. Where were the poets? I spent my free time in the North Dunedin Cemetery, sitting on the tombstones, dreaming, or walking along St Clair beach, by the lupins and the sandhills. One day I wrote a story, sent it to the *Listener*, and to my surprise it was published, with illustrations. I was very proud of my story. Though it was published only under my initials, someone traced it to me and added to my self-esteem by remarking, — The *Listener*'s hard to get into. Secretly, I was more proud of the illustration by Russell Clark than I was of the story, for the scene depicted a dark-haired attractive schoolgirl in the kitchen of a typical New Zealand home, such a contrast to the home I'd been describing, my own, which had never been 'typical', which had none of the furniture and ornaments portrayed by the artist. The schoolgirl in the smart gym frock was a flattering picture of myself. If I had been able to draw I would have made a 'true' picture, far from the pleasantly romantic etching.

As it was becoming impossible for me to reconcile 'this' and 'that' world, I decided to choose 'that' world, and one day when the Inspector was visiting my class at school I said, — Excuse me, and walked from the room and the school, from 'this' world to 'that' world where I have stayed, and where I live now. At first, it seemed a lonely disastrous choice. I tried to kill myself, and was sent to hospital for six weeks, and when I came out I found a living-in job as a housemaid in a boarding-house for workmen and old ladies, and in the evenings, in my small room which was still used as the linen cupboard though it now held my bed and chest of drawers squeezed in somehow beneath the shelves piled with linen, I sat on my bed with my newly acquired secondhand typewriter, an aged Barlock whose keys performed a roundabout dance before they reached the paper, and typed slowly, with one finger, for I had never used a typewriter before, most of my *Lagoon* stories.

Then, after another family tragedy, I went to live in Christchurch where I worked at a 'racing' hotel and continued writing the stories which I gave, as they were written, to a friend who had tried to make 'this' world more endurable for me. Though I tried to be a good waitress, my fear of everyone — the fierce black-clad head waitress, the proprietor, his wife, the guests, a fear intensified in some way by the nightmare of the long long Christchurch streets, became more than I could bear alone, and I was again sent to hospital.

For eight of the next ten years I was in hospital, and when once again I was discharged 'on probation' I was truly an inhabitant of 'that' world. Once again, optimistically I found a living-in job as a waitress in a Dunedin hotel where I saved enough wages to replace the old Barlock with a more modern secondhand Remington on which I typed stories and poems to send to the *Listener*. (I had pleasant memories of the publication of my first story.) Yet the pattern of fear began to reassert itself (I developed a terror of calling out the meal-orders to the sharp-tongued second cook — and how could I be a waitress if I could not give orders without bursting into tears?) and I returned to hospital for a short time. When I was discharged I knew (though I was repeatedly urged to 'adapt' 'mix' 'conform') that unless I devoted my time wholly to making designs from my dreams, whether or not those designs were approved or admired, I should spend the remainder of my life in hospital — or perhaps, as had been planned for me while I was a patient, there would be a leucotomy and the dreams of those who cared for me would be realized: I would 'mix' 'conform', become 'a useful member of the community'. I went to stay in Auckland where I met Frank Sargeson and accepted his invitation to live in the army hut in his garden. He arranged, through an understanding doctor, that I should receive a Social Security Benefit, for a time. I wrote *Owls Do Cry*. Then, because the doctor could not continue indefinitely to supply certificates, I was forced once again to venture into 'this' world, and spent a nightmare two weeks cleaning rooms at a huge hotel in Auckland, until I was sacked for my slowness — the housekeeper had found me, at four-thirty in the afternoon, still struggling to finish my allotted 'floor' when the other housemaids had been gone three and a half hours!

At the suggestion of Frank Sargeson I applied for a Literary Fund grant to travel overseas. I knew, and others knew, that leaving the country was my last hope to avoid life-long confinement in hospital, for the doctor who had supplied the certificates was already persuading himself into the way of thinking which I had met so often: nothing could be done for me, I would be 'better off' in hospital.

I lived in London, Ibiza, Andorra, and returned to London where I sought help at the Maudsley Hospital, and when the doctor had studied my history and difficulties, I was astonished and grateful to hear him refute all previous commandments. — Why mix, why conform? I think you need to write to survive. First write the story of your years in hospital, then keep on writing. You've no money, no income? We will arrange a National Assistance grant. There will be the usual personal difficulties, depressions; but we will try to make them more endurable.

During the four years that followed I wrote *Faces in the Water, The Edge of the Alphabet*, two volumes of stories, and *Scented Gardens for the Blind*, a novel-length autobiographical essay *Towards Another Summer* and other works.

The circumstances quoted by Virginia Woolf as being against writing have not changed. 'Dogs will bark; people will interrupt; money must be made; health will break down.' (The reference of the first has extended: people will play radios, televisions; traffic will change gear, jets will fly overhead.) One is lucky to get any writing done, particularly as there is a fifth circumstance which may be equally unfavourable: human beings will procrastinate, make excuses for beginning. A writer gets into the state of the condemned prisoner who, knowing that execution is inevitable, employs every device he can think of to prolong the walk to the scaffold; until words, like wardens, seize him at last. Here the similarity ends between the writer and the unhappy prisoner: the writer is punished after execution is carried out, and as it is he who pronounces his own sentence, so it is he who performs execution and punishment.

Though I began writing when I was a child and have never really stopped writing, I think I really began when my need to write was understood by distant members of the same profession which had previously condemned my needs as 'unhealthy', when I had enough money not to invite the fourth circumstance (waiting neighbour at any time) by trying to involve myself with waitressing; in short, when I had 'freedom' to write. Freedom to write is a very narrow freedom among the many personal imprisonments suffered by those who want to write, yet it is the master key, and if a writer has determination enough to turn the key (heedless of the desires and warnings of those who don't understand or fear (rightly) the consequences of this outrageous daylight robbery of the imagination) then he may be able to put his dreamed works into words. It is for the critics and psychologists to remind and explain the sad truth that the urge to write is not correlated with literary talent.

Though my personal and family life has seen many disasters, I've been lucky in the small, sometimes more influential circumstances of my life — meetings, partings. At least I've been lucky enough to have the services of a locksmith. The story of How I Began writing could be condensed into two lines:

A Hawk Came out of the Sky.
and
I Knew a man who knew a man who knew a locksmith.

beware : do not read this poem

tonite, thriller was
abt an ol woman, so vain she
surrounded herself w /
    many mirrors

it got so bad that finally she
locked herself indoors & her
whole life became the
    mirrors

one day the villagers broke
into her house , but she was too
swift for them . she disappeared
    into a mirror
each tenant who bought the house
after that , lost a loved one to

    the ol woman in the mirror :
    first a little girl
    then a young woman
    then the young woman/s husband

the hunger of this poem is legendary
it has taken in many victims
back off from this poem
it has drawn in yr feet
back off from this poem
it has drawn in yr legs

back off from this poem
it is a greedy mirror
you are into this poem . from
    the waist down
nobody can hear you can they ?
this poem has had you up to here
    belch
this poem aint got no manners
you cant call out frm this poem
relax now & go w / this poem

move & roll on to this poem
do not resist this poem
this poem has yr eyes
this poem has his head
this poem has his arms
this poem has his fingers
this poem has his fingertips

this poem is the reader & the
reader this poem

statistic: the us bureau of missing persons re-
            ports that in 1968 over 100,000 people
            disappeared leaving no solid clues
                  nor trace   only
        a space   in the lives of their friends

Ishmael Reed (b. 1938)

LAURIS EDMOND

Lauris Edmond, one of New Zealand's leading poets and most prolific and versatile writers, has gained wide recognition including the PEN-New Zealand Best First Book Award, the Katherine Mansfield Memorial Fellowship, the Commonwealth Poetry Prize, and an Honorary Doctorate of Literature from Massey University. At the award ceremony for this last she delivered the address *Imagining Ourselves*.

Imagining Ourselves

At its graduation ceremonies a university reaffirms its own faith in intellectual distinction and in the independence of mind it exists to foster, and of course at this as at other times it is watched and questioned by the community that has set it up, uses and pays for it, and must in the end shape its direction.

My own long association with this university has been as much from the outside as from the inside — more in fact. I have been one of your students but never a full-time or internal one. I have taught your courses but, except in vacations, always in Wellington not Palmerston North. I have learnt some of the most crucial lessons of my life in this university, particularly as a first time extramural student in the late Sixties. Then, rather late in the day — but there are many of us here tonight I think who are living proof that nothing is ever quite *too* late — I encountered the French Existentialists, read Jean-Paul Sartre and Camus and was, as millions before me had been, at first stunned and eventually transformed by this great Twentieth Century philosophy, which teaches a way of taking control of your own being and your own destiny. I have also made some of my closest friendships here, come to love, as everyone must, the seasonal changes in this beautiful campus, and shared in a small way in the shaping of new ways to think about English as a subject for university study. One way and another I have come to have an association with Massey University which has for me an enduring vitality and value.

There is one aspect of this experience that I want to talk about in a little more detail. I am a working writer, and although university English Departments exist to explore and illuminate literatures, their own and others, and the languages they use, for writers in the field there can be a conflict between their sense of their own work and of academic literary study. Writers long for recognition by the university, yet often affect to despise it. They can be ill-tempered about the 'ivory tower', its high salaries and what they see as a protected environment; they regard themselves as heroic and penurious battlers in the market place, while academics are elegant parasites.

I think it is equally likely — though I know less about this — that there is an image in the academic mind of vulgar literary self-promoters who drink too much, trumpet too loudly, are devoured by a belief in the permanent importance of their lightest word, and who, if you get caught with them, will bore you to extinction with stories about the philistines who don't understand their great works and, worse still, don't pay properly for them.

I would not presume to advise you about how much of these macabre portraits you should accept as authentic. I do however want to say a word about how, in my view, this university deals with the problem, real or imaginary. As sections of it have done distinguished pioneering work in Agricultural Science and many other fields, so at present the English Department in the Faculty of Humanities is taking important and innovative directions. This is a perfect moment for me to say so, because my sense of the originality, energy and democratic thrust of the only part of the university I really know adds immeasurably to my personal pleasure on this occasion.

Two aspects of this work interest me particularly. One is the reshaping of the basic introductory course in English Studies, so that it gives at once a world view and a proper local placing. Literature in this course is looked at as a development from one of the most fundamental needs we have as human beings, the need to imagine ourselves. Without this imaginative craving, and the capacity to satisfy it, we lack the means to enter into the full human inheritance. Similarly, as this course suggests, we cannot understand ourselves without knowing where our own literature began and how it came to be the truest image of ourselves we can have. New Zealand literature is seen as a colonial, or post-colonial, phenomenon, a cultural growth that parallels and defines our more general movement away from the English and European parent cultures, as well as coming to terms with the more precisely located Polynesian culture we also inherit. I don't think any other New Zealand university has approached the study of literature with quite this combination of the universal and the intimately focussed in one central course.

The other achievement, or perhaps I should call it a process, I want to mention is the exploration of New Zealand English in daily use that takes this English Department into the widest possible community. There is no question of the ivory tower here: English is studied as it exists and changes, within its conventions in business, in politics, in science, in personal communication. From story-telling to stock-taking, from dreaming to designing an engineering plant, language is our necessary tool, our challenge, sometimes our glory, often our defeat. How to make yourself understood on any one of the thousand different levels daily living requires — this, and not less, this course takes as its field of study and experiment.

My own work, the kind I like best to do, though by its nature it is rather intermittent and haphazard, is the writing of poetry. To say this is to make a tacit admission that I am in love with language, as of course I am. It has been a curious discovery for me that working in this course with the many ways of writing English, almost all prose, has given me a strong sense of the unity behind the infinite variety of English. It is a unity you can only find through a conscious exploration of the language as it serves many different purposes. If precision and exactness are central to the writing of a poem, so are they to the framing of a business memo, the working out of an application for a job, or — if you will forgive me — the writing of a speech. Some forms of language, like the cliché, dislocate our relationship with the language by separating the words we use and the action or thoughts they represent, moving them a long way apart. This again is as true for a poem or a play as it is for a pamphlet on the uses of electricity, or a discussion paper on a piece of new legislation — though of course as we all know the particular desire, or temptation, of politics is to use language for its capacity to tell lies rather than the truth.

There is a pleasurable harmony for me in these connections. However the forms of poetry may differ from those of prose, however reduced and intensified may be the world a poem gives, compared with the more leisurely and detailed visions of prose, we share the vocabulary, the nuances, of what is, to our immense good fortune, the richest and most various language on earth. Emily Dickinson, the great American 19th century writer, is speaking in the natural English of her day when she says:

> It was not Death, for I stood up,
> And all the Dead lie down —
> It was not night, for all the Bells
> Put out their Tongues, for Noon —[1]

James K. Baxter, too, uses natural New Zealand English, a little heightened, when he writes, in one of his later sonnets:

> Brother Ass, Brother Ass, you are full of fancies,
> You want this and that — a woman, a thistle,
>
> A poem, a coffeebreak, a white bed, no crabs:
> And now you complain of the weight of the Rider
>
> Who will set you free to gallop in the light of the sun! [2]

So do I, when I write, in one of my *Wellington Letter* poems:

> There are fixed points
> like stars; they wake each night
> after days of flux and we say —
> 'this is love'. It is not so easy —

to hold your frail poise
you must stand against me;
when the lout comes in to the room
you must leave and speak to him.[3]

We are all consciously using the words and cadences of English speech; not a special literary jargon, but words we hope will resonate in the minds of those we say them to, like the voice of a friend, someone they can trust.

There are other studies being carried out by teachers and scholars of English here, exciting in their erudition and innovative thrust to critics and readers well beyond New Zealand, and these equally I salute. But since my own experience of university study, and the experience of all the students I encounter, is of study inextricably mixed up with everyday living, it is the vital connection between Massey University and the prosaic events of jobs and families that best modifies that crude outline I gave you of an arrogant unrealistic academic world. Here at Massey I haven't found it exists.

But I want to say something too about the other subject of my stereotype, the working writer. Here too, I can only tell you of what I know. That requires me to tell you of my strange sense that I am not alone here, that I appear before you encompassed about by a shadowy company but one that is real and alive. In honouring me, you honour all the other women — and some men — who, like me, have not tried to distinguish between the importance of their children, their households, their close relationships, the jumble of their lives, and the books or poems or stories that had to be written — or indeed the essays that had to be in on time. Extramural study invites you to do this by bringing scholarly investigation to your kitchen table or a porch where you sit while a baby sleeps. The writing life, as I conceive it, does the same thing. Once I say to myself that life is one thing, art another, I believe I do irreparable damage to both. I have learnt not to try. The consequences are that my working life is cluttered, sometimes confused, marred by postponements and prevarications. I never think I have the life/art balance right, and I don't suppose I ever will. But this, I have come to think, is the only choice I have. It is within this muddled and amorphous framework — indeed at the heart of it — that I must look for the still centre from which writing comes. Katherine Mansfield, in a wonderful letter as yet unpublished, calls it a drawing upon 'one's real *familiar* life — to find the treasure in that'.

Women writers in New Zealand in earlier years did not, it seems to me, have as much choice as I do. Again and again, if you look at their history, they had to choose one or the other, to be a writer or to have a personal life that included children and family. I myself, I have to admit, have had to do them in tandem, life number two following life number one rather than completely coinciding with it.

There are many variations to this pattern. These others, predominantly women, but some men too, stand here with me because they have embraced this same principle, and work through the consequences of it as best they may, as I do. So do extramural students, so, in my experience of it, does this university. There is an aboriginal bark painting by Yirawala of the Gunwinggu tribe on Croker Island, West Arnhem Land in the Northern Territory of Australia, which shows Maralaitj, the creator-ancestress, giving birth to the northern tribes. It is in ochres on eucalyptus bark, and can be seen in the Australian National Gallery in Canberra. Maralaitj is shown standing front forward, her eyes wide and momentous-looking, her legs bent to help her support the weight of the northern tribes, which are to be seen — very small, very numerous — swimming downwards, through her body and further, through her birth canal and into the space below, where they continue on till they touch the ground. She is a splendid figure, resolute and capacious, and although the august administrators and teachers of this university might think this a highly dubious comparison, I have to tell you that it was this image of fruitfulness that first occurred to me when I came to consider what I might say of you here. I hope you will accept it as a truly respectful offering.

I am honoured by your recognition because I believe in the power of poetry to name our experience, to glimpse what it cannot name — the spiritual reverberations of everyday events — and sometimes to heal and restore what has been hurt or ruptured by experience. I am proud to be a woman in a new period of women's recognition of themselves. Most of all I acknowledge this honour as an affirmation of the profound interdependence that I believe exists, if we can find it, between the ordinary habits and events of our lives and the furthest reaches of artistic or scholarly discovery.

from *Little Gidding* V

What we call the beginning is often the end
And to make an end is to make a beginning.
The end is where we start from. And every phrase
And sentence that is right (where every word is at home,
Taking its place to support the others,
The word neither diffident nor ostentatious,
An easy commerce of the old and the new,
The common word exact without vulgarity,
The formal word precise but not pedantic,
The complete consort dancing together)
Every phrase and every sentence is an end and a beginning,
Every poem an epitaph. And any action
Is a step to the block, to the fire, down the sea's throat
Or to an illegible stone: and that is where we start.

T. S. Eliot (1888–1965)

NOTES

Anon. *Pangur Bán*

1 The writer was an Irish monk living in the early ninth century on the shores of the Bodensee (between what are now Switzerland and Germany). *Bán* means 'white'.

Aristotle

1 Athenian tragic poets. Little of Sthenelus' work, and nothing of Cleophon's, survives.

Chaucer

1 Geoffrey Chaucer, in this versified reproof to his scribe or copyist Adam, reveals some of the difficulties medieval writers laboured under.
2 Chaucer wrote *Boece* (his translation of Boethius' *Consolation of Philosophy*) and *Troilus and Criseyde* before he embarked upon his *Canterbury Tales*.
3 That is, unless you copy my compositions more accurately.
4 Any accidental errors had to be scraped and rubbed with knife and pumice from the surface of the parchment.

Caxton

1 Documents.
2 German.
3 Thames.
4 Inn.
5 Provincial.
6 Undertake.
7 Ovid (the Roman poet Publius Ovidius Naso, b. 43 B.C.).
8 Unlearned.
9 Learning.
10 John Skelton (d. 1529) poet, dramatist, and tutor to the future King Henry VIII.
11 Skilfully.
12 Future.

Sidney. *Defence of Poesie*

1 *poieten*.
2 *poiein*.
3 The suspension mark or tilde [~] above some letters is a space-saving device inherited by the early printers from their manuscript-writing predecessors. In this document it indicates *n* or *m*. Editors sometimes restore such letters in italics. (Cf. the extract from Caxton.)
4 Probably 'cunning' (= skill).

5 Theagenes is the hero of the 3rd century A.D. Greek romance *Aethiopica* by Heliodorus of Emesa; Pylades is the friend of Orestes in Greek mythology; Orlando is the Italian name of the Frankish warrior Roland, central figure of the *Chanson de Roland* and of various 15th and 16th century romantic epics including Ariosto's *Orlando Furioso*, which Sidney probably has in mind here; Cyrus the Great is the idealised central character of the *Cyropaedia* of Xenophon (5th century B.C.); and Aeneas is the eponymous hero of Virgil's epic poem.

6 *mimesis.*

7 A popular romance circulated in various European languages in the 15th and 16th centuries.

8 'Shall this earth see me running away? Is it such wretchedness to die?' Virgil, *Aeneid*, XII, 645–6.

9 *The Consolation of Philosophy* of Boethius (d. A.D. 525) uses verse and prose to expound Platonic ideas.

10 'To follow their impulses.'

11 Cf. The Knight's Tale, *Canterbury Tales*, I(A) 886.

12 'To beg the question.'

13 The ferryman of the dead over the river Styx.

14 See 2 Samuel 12.

15 *eikastike.*

16 *phantastike.*

17 Rampart.

18 'I bid him cheerfully remain a fool' (Horace, *Satires*, I, i, 63).

19 Terms used in philosophy.

20 Construe, explain.

21 'Which authority certain barbarous and uncultivated persons wish to use for expelling poets from the state.' Cf. Scaliger, *Poetics*, I, 2.

Bacon. *Advancement of Learning*

1 'Accursed is this mob, which does not know the law' (John 7:49).

2 'Ten years have I spent reading Cicero.' 'You ass.' Cf. Erasmus, *Colloquies*: The Young Man and the Echo.

3 'To a greater or lesser degree.'

4 That is, heavily decorated. Elizabethan obsession with manner rather than matter reached its extreme in the style known as *euphuism*, after John Lyly's *Euphues*, of which the following is a sample:
' . . . Ah Livia, Livia, thy courtly grace without coynes, thy blazing beauty without blemish, thy courteous demeanour without curiosity, thy sweet speach savoured with wit, thy comly mirth tempered with modesty, thy chast looks yet lovely, thy sharp taunts yet pleasant, have given me such a checke, that sure I am at the next view of thy vertues, I shall take thee mate: And taking is not of a pown, but of a prince, the losse is to be accompted the lesse. And though they be commonly in a great choler that receive the mate, yet would I willingly take every minute ten mates, to enjoy Livia for my loving mate. Doubtlesse if ever she hir self have been scorched with the flames of desire, she will be ready to quench the coales with courtesie in an other, if ever shee have been attached of love, she will rescue the feaver of fancie, she wil help his ague, who by a quotidian fit is converted phrensie: Neither can there bee under so delicate a hew lodged deceite, neither in so beautiful a mould a malicious minde.'

5 Concerned with transferring.

6 'Concerning the "notes", or marks, of things.'

7 'Arbitrary.'
8 'Impreses' are heraldic crests and coats of arms; 'emblems' are pictures with a moral meaning.
9 For the first and second general curses see Genesis 3: 16–19 and 11: 6–8.
10 'In a scattered fashion.'
11 'I would prefer a course of our meal to have pleased the guests rather than the cooks.' Cf. Martial, ix, 82.
12 'What seems old is very out of place if new.'
13 Written symbols.
14 'All things by all means.'

Bacon. *Studies*

1 'Studies become manners.'
2 'Hair-splitters' (literally 'dividers of cumin-seed').

Milton. *Areopagitica*

1 Conduct.
2 Ovid (*Metamorphoses*, III and VII) tells two stories in which dragons' teeth, sown like seed, spring up as armed men.
3 The world was thought to consist of four essences or elements (earth, air, fire and water), and the heavens of a fifth element known as ether or the 'quintessence.'
4 An ecclesiastical tribunal operative from the 13th century onwards in Germany, France, Italy, and (notably) Spain.
5 Cf. Exodus 16: 16ff.
6 Cf. Matthew 15: 17-20; Mark 7: 14–23.
7 Ecclesiastes 12: 12.
8 Acts 19: 19.
9 Venus, angry at the love between her son Cupid and the mortal girl Psyche, gave the latter the task of separating mixed seeds (cf. Apuleius, *The Golden Ass*, IV–VI).
10 Cf. Genesis 3.
11 Superficial.
12 Milton cites St Thomas Aquinas (d. 1274) and Duns Scotus (d. 1308) as representatives of medieval theology and philosophy.
13 Cf. Spenser, *Faerie Queene*, II, vii and xii. The Palmer does not in fact accompany Sir Guyon into the Cave of Mammon.
14 That is, if we have escaped the school cane only to come under the discipline of the censor.
15 Published.
16 Of Pallas Athene, and therefore of wisdom.
17 Youth, minor (French *puis-né*, after born).
18 *An Advertisement Touching the Controversies of the Church of England.*
19 John Knox (1505–71) played a leading part in the Reformation, especially in Scotland.
20 Milton probably got this story and its interpretation from Plutarch's *On Isis and Osiris.*
21 Appearing to pass close to the sun.
22 Leading church reformers.
23 Doctrinal system.

24 Of the same kind throughout.
25 That is, the parts combine harmoniously to make the whole.

Milton. *When I consider . . .*

1 Cf. the Parable of the Talents (Matthew 25: 14–30). Milton is referring to his talent (in the more modern sense) as a writer, rendered 'useless' by his blindness.

Locke

1 *verbi gratia* ('for the sake of the word'). Compare with 'e.g.,' *exempli gratia.*

Dryden

1 Nisus lost a race to his friend Euryalus by slipping in the blood of a sacrificed animal (cf. Vergil *Aeneid*, V, 315ff.).
2 Controlled versification.
3 The youthful heir to the Emperor Augustus. Vergil (*Aeneid*, VI, 860ff.) mourns his death at the age of twenty.

Swift. *Tatler*

1 A pseudonym first used by Swift for an attack on John Partridge, an almanac-maker and quack astrologer, in 1708.
2 'Originally the name of a street near Moorfields in London, much inhabited by writers of small histories, dictionaries, and temporary poems, whence any mean production is called *grubstreet*' (Johnson's *Dictionary*).
3 In the letter (no doubt a fictitious product) Swift italicises the terms and phrases he deplores.
4 Abbreviations for *physiognomy, hypochondria, mobile vulgus, positive, reputation.*
5 Originally, a list of passages to be deleted before a book was cleared by the Roman Catholic church for publication.
6 'Simple elegance.' Cf. Horace, *Odes*, I, v, 5.

Swift. *On the Death of Dr Swift*

1 Four letters, written by Swift under the pseudonym of a Dublin draper (or drapier), which attacked Prime Minister Walpole ('great Sir Robert') and English profiteering in Ireland.

Johnson. *Preface to the Dictionary*

1 Hooker actually wrote: 'True withal it is, that alteration, though it be from worse to better, hath in it inconveniences, and those weighty.' Cf. *Of the Laws of Ecclesiastical Politie*, IV, 14.
2 See Spenser, *The Faerie Queene*, IV, ii, 32.
3 'A Proposal for Correcting, Improving and Ascertaining the English Tongue' (1712).

Pope

1 Cf. Ben Jonson, *Every Man Out Of His Humour.*

2 The fabled home of the Muses.
3 A line of twelve syllables, as illustrated by Pope in the next line.
4 Two 17th-century English poets.
5 Cf. Dryden, *Alexander's Feast.*

Johnson. *Life of Cowley*

1 Aristotle.
2 *techne mimetike.*
3 See p. 55 above.
4 'Dissonant harmony.'
5 The Athenian philosopher Epicurus (341-270 B.C.) saw life as mechanistic rather than divinely ordained. His gods therefore had little to do.
6 That is, thinness of constituent parts.

Cowper

1 A Biblical sea-monster. See Wilde, n. 23 below.

Gibbon

1 Publius Cornelius Tacitus (*ca.* A.D. 55–117), Roman historian. Gaius Plinius Caecilius Secundus (*ca.* A.D. 61–113) wrote widely but is best remembered for his nine books of 'Letters'. Decimus Junius Juvenalis (b. *ca.* A.D. 65), poet and satirist.
2 Cassius Dio Cocceianus (*ca.* A.D. 150–235) wrote, in Greek, a 'Roman History' in eighty books. Ammianus Marcellinus (b. *ca.* A.D. 330) wrote a continuation of Tacitus' Histories.
3 Louis le Nain Tillemont (1637–98) wrote histories of Rome and of the Christian Church up to the 7th century.
4 Lodovico Antonio Muratori (1672–1750), Carlo Sigonio (1520–84), Francisco Scipio marchese di Maffei (1675–1755), Cesare Baronio (1538–1607) and Antoine Pagi (1624–99) wrote, in Latin, on the history of Italy.
5 In *Decline and Fall* (ii, 319), Gibbon describes James Godefroy (1587–1652) as 'balanced by the opposite prejudices of a Civilian and a Protestant.'
6 Nathaniel Lardner (1684–1768) pioneered research into early Christian literature.
7 'To read intensively rather than widely.'
8 Sir William Blackstone (1723–80) published his *Commentaries on the Laws of England* in the years 1675–79.
9 William Law (1686–1761) is best known for his *Serious Call to a Devout and Holy Life.*
10 David Hume (1711–76), one of the most influential Scottish philosophers of the 18th century, also wrote on history.

Boswell

1 Boswell had been legal counsel for a Scottish schoolmaster sacked for brutality.

Byron

1 The supposed author of the 1st century A.D. Latin treatise *On the Sublime.*

Hazlitt

1 Sterne, *Tristram Shandy*, III, xx; Shakespeare, *Hamlet*, II, ii (free quotation).
2 *A Grammar of the English Language*, 1818, Letter xxiii.
3 See Johnson, *Preface to the Dictionary*, n. 2 above.
4 Shakespeare, *Hamlet*, II, ii.
5 Anthology (literally 'a gathering of flowers').
6 Craze for tulips.
7 'Speech creeping on the ground' (a possible reference to Horace, *Epistles*, II, i, 250–1: *sermones . . . repentes per humum*).
8 Shakespeare, *Julius Caesar*, II, i (free quotation).
9 *The Winter's Tale*, IV, iii.
10 'Fantoccini': marionettes, puppets. The quotation is from Shakespeare, *Macbeth*, V, v.
11 Cf. Milton, *Paradise Lost*, IV, 988–9: 'and on his crest Sat Horror plumed.'
12 Shakespeare, *Twelfth Night*, I, v.
13 Cowper, *The Task*, V, 173-6.

Borrow

1 Named after.
2 Thomas Blood, a 17th-century adventurer.
3 Borrow had been employed by publisher Sir Richard Phillips in updating the *Newgate Calendar*, a popular account of notorious crimes.
4 Assistant to Phillips (see n. 3).
5 Borrow had talked with a 'pea-and-thimble man' shortly before this episode.

Dickinson

1 The numbering of the poems is that established by Thomas H. Johnson in his *Poems of Emily Dickinson* (1951).

Ward

1 Mrs Ward had translated the remarkable diary published in 1883 by Swiss author Henri-Frédéric Amiel (1821–81).
2 Mrs Ward's first novel, published in 1884.
3 Cf. Kipling's 'In the Neolithic Age' (1895):
      There are nine and sixty ways of constructing tribal lays,
      And every single one of them is right!

Hopkins

1 Robert Bridges (1844–1930), Hopkins's friend and editor.

Wilde

1 The names of Wilde's two sons.
2 Cf. Wordsworth, *Power of Music*, l. 4.
3 Leontini was an ancient Greek settlement in Sicily, known for its philosophers.
4 Official Parliamentary report.

5 'Corner of the world.' Both French phrases are borrowed from the French novelist Emile Zola (1840–1902).

6 Sir Francis Burnand (1836–1917) was editor of *Punch* and wrote a number of burlesques for the stage.

7 Herbert Spencer (1820–1903) developed the doctrine of Evolution into the philosophical idea of a Life Force.

8 Cf. Shakespeare, *The Tempest*, *A Midsummer Night's Dream*, and *Macbeth*; also Wordsworth's sonnet, 'The world is too much with us'.

9 Shakespeare, *Hamlet*, III, ii.

10 Cf. Shakespeare, *Othello*.

11 Shelley, *Prometheus Unbound*, I, 748.

12 Cf. Shakespeare, *2 Henry IV*, III, i.

13 18th-century highwaymen.

14 Arthur Schopenhauer (1788–1860) saw the world as a malignant illusion.

15 Ivan Turgenev (1818–83) portrayed Nihilism, a revolutionary terrorist movement of tsarist Russia, in his novel *Fathers and Sons*.

16 Robespierre (1758–94), one of the leaders of the French Revolution, based his actions on the writings of philosopher Jean-Jacques Rousseau (1712–78). In 1887 Queen Victoria opened the People's Palace in Mile End Road, London, for the enjoyment of the poor. The idea had occurred in Walter Besant's novel *All Sorts and Conditions of Men*.

17 Rolla, the eponymous hero of a poem by French poet Alfred de Musset (1810–57), is redeemed from debauchery by love. In Goethe's romance, *The Sorrows of Young Werther*, the hero commits suicide for love.

18 Cf. Homer, *Odyssey*, XIII (Morris's translation, l. 295).

19 Cf. Euripides, *Ion* (*ca.* 417 B.C.) and Horace, *Odes*, III, xi, 35.

20 Francisco Sanchez wrote *A Treatise on the Noble and High Science of Nescience* (1581).

21 The olive grove near Athens where Plato taught was named the Academy after its owner Academius.

22 See *The Temptation of St Antony*, vii.

23 Biblical monsters and symbols of Chaos, sometimes identified as whale, crocodile, or hippopotamus.

24 See n. 4 above. The Second Empire lasted in France from 1852 to 1870.

25 The quotations come from Tennyson's *The Princess* and from William Blake's 'The Evening Star' respectively.

Kipling. *A Legend of Truth*

1 'What is truth?' (John 18: 38).

Kipling. *Working Tools*

1 Kipling is parodying his own imperialist poem 'Recessional.'

2 'Believe one who has proved it.'

3 Cf. Acts 5: 1–11.

4 Cf. Acts 22: 28.

5 C —— had been Kipling's English and Classics Master at the United Services College at Bideford in Devon: '. . . a rowing-man of splendid physique, and a scholar who lived in secret hope of translating Theocritus worthily. He had a violent temper, no disadvantage in handling boys used to direct speech, and a

gift of schoolmaster's "sarcasm" which must have been a relief to him and was certainly a treasure-trove to me. Also he was a good and House-proud House-master. Under him I came to feel that words could be used as weapons, for he did me the honour to talk at me plentifully; and our year-in year-out form-room bickerings gave us both something to play with. One learns more from a good scholar in a rage than from a score of lucid and laborious drudges; and to be made the butt of one's companions in full form is no bad preparation for later experiences' (*Something of Myself*, ch. 2).

6 In Chaucer's *Canterbury Tales* it is not the Wife of Bath but the Cook who has a 'mormal' (running sore) on the leg.

7 Cf. 1 Samuel 28.

8 Cf. 1 Chronicles 11: 15–19.

9 The Abbé Prévost (1697–1763) published his notorious novel *Manon Lescaut* in 1731. Scarron (1610–60) published *Le Roman Comique* in 1651.

Frost

1 Zeno of Elea, who was born *ca.* 490 B.C., argued that a flying arrow is, at any given moment, stationary.

2 Niels Bohr (1885–1962) won the Nobel Prize for Physics in 1922.

Moore

1 Marianne Moore later reduced this poem to three lines:
    I, too, dislike it.
    Reading it, however, with a perfect contempt for it, one discovers in it, after all, a place for the genuine.

Woolf

1 Virginia Woolf's husband Leonard.

2 Geraldine Jewsbury (1812–80) published several novels.

3 Published in 1929 as *A Room of One's Own*.

4 Characters in Virgil's *Aeneid* and Homer's *Odyssey* respectively.

5 Cf. in Gilbert and Sullivan's opera *Patience* the poet Bunthorne's ambition to become:
                A pallid and thin young man,
                A haggard and lank young man,
            A greenery-yallery, Grosvenor Gallery,
                Foot-in-the-grave young man!

6 House of Vanessa and her husband Clive.

7 Lytton Strachey, Roger Fry, and E. M. Forster.

8 Lesbian.

9 A dog.

10 Hugh Walpole.

11 Eventually *Three Guineas*.

12 Elly Rendell, Virginia Woolf's doctor.

13 Virginia Woolf's brother.

14 The word is illegible.

15 G. W. Rylands.

Pound. *Ode pour l'Election de son Sépulchre*

1 One of the 'seven against Thebes' of classical mythology, killed by a thunderbolt after defying the gods.
2 In these words the Sirens, trying to lure Odysseus to destruction, tell him they know all about his past. Cf. Homer, *Odyssey*, XII, 184ff.
3 Odysseus' faithful wife, to whom he returns in the course of the *Odyssey*.
4 French novelist (1821–80), author of *Madame Bovary*, etc.
5 The enchantress who delayed Odysseus' return home for a year, and who suggested to him how to hear the Sirens with 'unstopped ear' and survive.
6 'In his thirtieth year.' Pound is quoting the 15th-century French poet François Villon.

Pound. *ABC of Reading*

1 These terms name the three functions of language Pound has just listed.
2 Stanza.

Mansfield

1 'You know what I mean.'
2 Cf. Shakespeare, *The Tempest*, III, ii.
3 'Flying' — an allusion to the legend of the Flying Dutchman, a ship doomed for ever to search for harbour and celebrated in Wagner's opera, *Der Fliegende Holländer*.

Orwell. *Why I Write*

1 Field Marshal Lord Kitchener (1850–1916), after a distinguished if stormy military and political career, was drowned when the ship carrying him on a mission to Russia struck a mine.
2 'Occasional verses' — i.e., for a particular occasion.
3 Milton, *Paradise Lost*, II, 1021–2.
4 Eugene Aram, a Yorkshire schoolmaster forced by poverty to consent to a murder, was executed in 1759.
5 The opening line of a popular song by Alfred Bunn (1796–1860):
> I dreamt that I dwelt in marble halls,
> With vassals and serfs at my side . . .
6 Orwell's next book was to be *Nineteen Eighty Four*.

Koestler

1 'Behold the man' (John 19: 5).
2 St Augustine of Hippo (354–430) described his early profligacy and subsequent conversion to Christianity in his *Confessions*; Jean-Jacques Rousseau (1712–78) recounted his unstable and unhappy life in his posthumously published *Confessions*; and Thomas de Quincey (1785–1859) published *Confessions of an English Opium Eater* in 1822.
3 Benvenuto Cellini (1500–71) wrote vividly of the political and artistic life of his day.

Betjeman

1 John Murray, the publishing firm.
2 The church of St Mary Redcliffe, Bristol.

Frye

1 Samuel Butler (1835–1902) wrote, often satirically, upon a wide variety of subjects. He spent five years as a sheep-breeder in New Zealand.
2 'We've changed all that' (Molière, *Le Médecin Malgré Lui*, II, iv).
3 *The Defence of Poesie* — see p. 17 above.
4 Cf. Hopkins's letter to Robert Bridges, 25 September 1888.
5 From *An Ordinary Evening in New Haven*, XXVIII.
6 Cf. *The Advancement of Learning*, II, iv, 2.
7 The motto on the title-page of Yeats's *Responsibilities* (1914) reads: '*In dreams begins responsibility*' — *Old Play*.

Lessing

1 A science-fiction novel published in 1930 by Olaf Stapledon (1886–1950).
2 *The Popol Vuh* contains the legends and history of the Quiché Indians of Guatemala; the Dogon people of Mali possess a sophisticated cosmology and system of metaphysics; Gilgamesh is the hero of ancient Sumerian epic.

Auden

1 In August 1968, Czechoslovakia was invaded by the combined armed forces of its 'allies' of the Warsaw Pact, and at the Democratic Party convention in Chicago police brutality received official sanction.

Ihimaera

1 'Tangi': a Maori funeral.

Frame

1 A reference to Sir Philip Sidney's *The Defence of Poesie*. See p. 11 above.

Edmond

1 *Poems of Emily Dickinson*, ed. Thomas H. Johnson (Belknap Press, Harvard University Press), no. 510, p. 391.
2 *Collected Poems*, ed. J. E. Weir (Oxford University Press, 1979), p. 472.
3 *Selected Poems* (Oxford University Press, 1984), p. 50.

SOURCES AND ACKNOWLEDGEMENTS

Grateful acknowledgement is made to the publishers and copyright owners of the following:

Anon., 'I and Pangur Bán my Cat': from Robin Flower, *The Irish Tradition*, London, Oxford University Press, 1947.

Aristotle: from *Aristotle On the Art of Poetry*, translated by Ingram Bywater, Oxford, Clarendon Press, 1920.

Auden: from W. H. Auden, *Collected Poems*, edited by Edward Mendelson, London, Faber and Faber, 1976. Reprinted by permission of Random House, Inc., Alfred A. Knopf, Inc.

Austen: from *The Novels of Jane Austen*, edited by R. W. Chapman, Vol. V., Oxford, Clarendon Press, 1933.

Bacon, 'Of Studies': from Sir Francis Bacon, *The Essayes or Counsels*, edited by Michael Kiernan, Oxford, Clarendon Press, 1985.

Bacon, 'The Advancement of Learning': from Francis Bacon, *The Advancement of Learning and New Atlantis*, edited by Arthur Johnston, Oxford, Clarendon Press, 1974.

Betjeman: from John Betjeman, *Collected Poems*, London, John Murray, 1970.

Blake: from William Blake, *Works*, edited by John Sampson, London, Oxford University Press, 1913.

Borrow: from *The Works of George Borrow*, edited by Clement Shorter, Vol. 4, London, Constable, 1924.

Boswell: from *Boswell's Life of Johnson*, edited by G. B. Hill & L. F. Powell, London, Oxford University Press, 1952.

E. B. Browning: from *The Oxford Book of Victorian Verse*, chosen by Arthur Quiller-Couch, London, Oxford University Press, 1912.

Byron: from George Gordon, Lord Byron, *Poetical Works*, London, Oxford University Press, 1945.

Caxton: from *Caxton's Eneydos*, edited by M. T. Culley and F. J. Furnivall, London, Published for the Early English Text Society by Kegan Paul, Trench and Trübner, 1890 (EETS Extra Series 57). With acknowledgement to the Council of the Early English Text Society.

Chaucer: from *The Works of Geoffrey Chaucer*, edited by F. N. Robinson, 2nd ed., Boston, Houghton Mifflin, 1957. Copyright © 1957 by Houghton Mifflin Company.

Cowper: from *Life and Works of William Cowper*, edited by R. Southey, London, Baldwin & Cradock, 1835–37.

Cummings, 'since feeling is first': reprinted from *is 5* poems by E. E. Cummings, edited by George James Firmage, by permission of Liveright Publishing Corporation. Copyright © 1985 by E. E. Cummings Trust. Copyright 1926 by Horace Liveright. Copyright © 1954 by E. E. Cummings. Copyright © 1985 by George James Firmage. Also with acknowledgement to Harper Collins Publishers.

Dickinson, 'Shall I take thee', 'A word is dead', 'There is no Frigate like a Book', 'Could mortal lip divine', 'Your thoughts don't have words every day', 'It was not death': reprinted by permission of the publishers and the Trustees of

Amherst College from *The Poems of Emily Dickinson*, edited by Thomas H. Johnson, Cambridge, Mass., The Belknap Press of Harvard University Press, 1983. Copyright © 1951, 1955, 1979, 1983 by the President and Fellows of Harvard College.

Dryden: from John Dryden, *Poems*, edited by John Sargeaunt, Oxford, Clarendon Press, 1910.

Edmond: from Lauris Edmond, *Imagining Ourselves*, Palmerston North, Massey University, 1988. Reprinted by kind permission of the author.

Eliot: from *Four Quartets*, © 1943 by T. S. Eliot, New York, Harcourt, Brace & World, Inc. With acknowledgement to Harcourt Brace Jovanovich Inc. Also in *The Complete Poems and Plays of T. S. Eliot*, London, Faber and Faber, 1969.

Frame: from *Beginnings: New Zealand Writers Tell How They Began Writing*, edited by Robin Dudding, Wellington, Oxford University Press, 1980. Reprinted by permission of Curtis Brown (Aust.) Pty Ltd., Sydney.

Frost, 'A Considerable Speck (Microscopic)': from *The Poetry of Robert Frost*, edited by Edward Connery Lathem. Copyright © 1969 by Holt, Rinehart and Winston. Copyright 1942 by Robert Frost. Copyright © 1970 by Lesley Frost Ballantine. Reprinted by permission of Henry Holt and Company, Inc., and with acknowledgement to Jonathan Cape.

Frost, 'Education by Poetry': from *Selected Prose of Robert Frost*, edited by Hyde Cox and Edward Connery Lathem. Copyright © 1966 by Holt, Rinehart and Winston. Reprinted by permission of Henry Holt and Company, Inc.

Frye: from Northrop Frye, *Divisions on a Ground: Essays in Canadian Culture*, edited by James Polk, Toronto, House of Anansi Press Ltd, 1982. Reproduced by kind permission of the author.

Gibbon: from *Gibbon's Autobiography*, edited by M. M. Reese, London, Routledge and Kegan Paul, 1971.

Graves: from Robert Graves, *Collected Poems 1975* , London, Cassell, 1975. Reprinted by permission of A. P. Watt Ltd on behalf of the Trustees of the Robert Graves Copyright Trust.

Hazlitt: from *The Complete Works of William Hazlitt*, edited by P. P. Howe, London, J. M. Dent & Sons, 1931.

Herrick: from *The Poetical Works of Robert Herrick*, edited by F. W. Moorman, London, Oxford University Press, 1921.

Hopkins: from *Poems of Gerard Manley Hopkins*, 3rd ed., edited by W. H. Gardner, London, Oxford University Press, 1948.

Hughes: from Ted Hughes, *The Hawk in the Rain*, London, Faber and Faber Ltd.

Ihimaera: from Witi Ihimaera, 'Why I Write', *World Literature Written in English*, Vol 14, No 1 (1975). Reprinted by permission of Richards Literary Agency, Auckland.

Johnson: from *Samuel Johnson*, edited by Donald Greene, Oxford, Oxford University Press, 1984.

Kipling, 'A Legend of Truth': from Rudyard Kipling, *Debits and Credits*. Copyright 1926 by Rudyard Kipling. Used by permission of Doubleday, a division of Bantam Doubleday Dell Publishing Group, Inc.

Kipling, 'Working Tools': from Rudyard Kipling, *Something of Myself*, New York, Doubleday, 1937. Copyright 1937 by Caroline Kipling. Used by permission of Doubleday, a division of Bantam Doubleday Dell Publishing Group, Inc.

Koestler: from Arthur Koestler, *Arrow in the Blue*, London, Hutchinson 1952 (1969). Reprinted by permission of the Peters Fraser, & Dunlop Group Ltd.

Lessing: from Doris Lessing, *Canopus in Argos: Archives. Re: Colonised Planet 5*

Shikasta, London, Jonathan Cape, 1979. Reprinted by permission of Alfred A. Knopf, Inc.

Locke: from John Locke, *An Essay Concerning Human Understanding*, edited by Peter H. Nidditch, Oxford, Clarendon Press, 1975.

MacLeish: from Archibald MacLeish, *Collected Poems*, © 1917, 1952 by Houghton Mifflin Company.

Mansfield: from *The Journal of Katherine Mansfield*, edited by J. Middleton Murry, London, Constable, 1927.

Milton, 'Sonnet XVII': from *The Complete Works of John Milton*, edited by H. C. Beeching, London, Oxford University Press, 1941.

Milton, 'Areopagitica': from *The Complete Prose Works of John Milton*, Vol. II, edited by E. Sirluck, New Haven, Yale University Press, 1959.

Moore, 'Poetry': reprinted with permission of Macmillan Publishing Company from *Collected Poems of Marianne Moore*. Copyright 1935 by Marianne Moore, renewed 1963 by Marianne Moore and T. S. Eliot. Also in Marianne Moore, *Complete Poems*, London, Faber and Faber Ltd.

Orwell: 'Why I Write' from George Orwell, *England Your England and Other Essays*, London, Secker & Warburg 1954; 'Politics and the English Language', copyright 1946 by Sonia Brownell Orwell and renewed 1974 by Sonia Orwell, reprinted from *Shooting an Elephant and Other Stories* by George Orwell, reprinted by permission of Harcourt Brace Jovanovich, Inc. and of A. M. Heath and Company Ltd.

Pope: from Alexander Pope, *Poetical Works*, edited by Herbert Davis, London, Oxford University Press, 1966.

Pound, 'E.P. Ode pour l'Election de son Sépulchre': from Ezra Pound, *Personae*. Copyright 1926 by Ezra Pound. Reprinted by permission of New Directions Publishing Corporation. Also in *Collected Shorter Poems*, London, Faber and Faber Ltd.

Pound: from *ABC of Reading*. Reprinted by permission of New Directions Publishing Corporation, and with acknowledgement to Faber and Faber Ltd.

Reed, 'beware, do not read this poem' © 1972 Ishmael Reed. Reprinted by permission.

Shakespeare: from *Shakespeare's Sonnets*, edited by M. Seymour-Smith, London, Heinemann 1963

Sidney, 'Loving in Truth': from *Poems of Sir Philip Sidney*, London, Oxford University Press, 1970.

Sidney, 'Defence of Poesie': from *The Prose Works of Sir Philip Sidney*, edited by A. Feuillerat, London, Cambridge University Press, 1962.

Spender: from Stephen Spender, *Collected Poems*, London, Faber and Faber, 1955. Reprinted by permission of Random House, Inc., Alfred A. Knopf, Inc.

Spenser: from *Works of Spenser*, edited by R. Morris, London, Macmillan, 1897.

Swift: from Jonathan Swift, *Satires and Personal Writings*, edited by W. A. Eddy, London, Oxford University Press, 1932.

Thomas: from Edward Thomas, *Collected Poems*, Oxford, Clarendon Press, 1978.

Ward: from Mrs Humphry Ward, *A Writer's Recollections*, London, Collins, 1918.

Wilde: from *Oscar Wilde*, edited by Isobel Murray, Oxford, Oxford University Press, 1989.

Woolf: from Virginia Woolf, *A Writer's Diary*, copyright 1954 by Leonard Woolf and renewed 1982 by Quentin Bell and Angelica Garnett, reprinted by permission of Harcourt Brace Jovanovich, Inc.

Wordsworth: from William Wordsworth, *Poetical Works*, edited by Thomas Hutchinson, London, Oxford University Press, 1950.

INDEX OF AUTHORS AND TITLES,
AND OF FIRST LINES OF POEMS

My poem's epic, and is meant to be   77